Intelligent Systems Reference Library

Volume 265

The aim of this series is to publish a Reference Library, including novel advances and developments in all aspects of Intelligent Systems in an easily accessible and well structured form. The series includes reference works, handbooks, compendia, textbooks, well-structured monographs, dictionaries, and encyclopedias. It contains well integrated knowledge and current information in the field of Intelligent Systems. The series covers the theory, applications, and design methods of Intelligent Systems. Virtually all disciplines such as engineering, computer science, avionics, business, e-commerce, environment, healthcare, physics and life science are included. The list of topics spans all the areas of modern intelligent systems such as: Ambient intelligence, Computational intelligence, Social intelligence, Computational neuroscience, Artificial life, Virtual society, Cognitive systems, DNA and immunity-based systems, e-Learning and teaching, Human-centred computing and Machine ethics, Intelligent control, Intelligent data analysis, Knowledge-based paradigms, Knowledge management, Intelligent agents, Intelligent decision making, Intelligent network security, Interactive entertainment, Learning paradigms, Recommender systems, Robotics and Mechatronics including human-machine teaming, Self-organizing and adaptive systems, Soft computing including Neural systems, Fuzzy systems, Evolutionary computing and the Fusion of these paradigms, Perception and Vision, Web intelligence and Multimedia.

Indexed by SCOPUS, DBLP, zbMATH, SCImago.

All books published in the series are submitted for consideration in Web of Science.

Markus Mueck · Christophe Gaie
Editors

European Digital Regulations

Springer

Editors
Markus Mueck
Intel Labs
Unterhaching, Germany

Christophe Gaie
Centre de l'enseignement militaire
supérieur-Terre (CEMS-T)
Engineering and Digital Innovation
Division at the Prime Minister's Office
Paris, France

ISSN 1868-4394 ISSN 1868-4408 (electronic)
Intelligent Systems Reference Library
ISBN 978-3-031-80808-1 ISBN 978-3-031-80809-8 (eBook)
https://doi.org/10.1007/978-3-031-80809-8

Disclaimer: Within the present book, extensive reference is given to applicable European Regulations and explanatory information available on European Commission web sites. Key information is extracted and included in the relevant chapters for better readability. Please also note that this document is intended to share the authors' understanding of the current regulatory landscape of the European Union's digital regulations. It is not intended to constitute legal advice or lawful standing.

This Springer imprint is published by the registered company Springer Nature Switzerland AG
The registered company address is: Gewerbestrasse 11, 6330 Cham, Switzerland

Foreword

Digital technologies have undergone radical changes in the last decades and are currently being further developed, generating new opportunities for businesses and society. However, some of these digital technologies will have ambivalent consequences, which cannot be assessed by today. Following the tradition of the regulatory approach within the European Union, i.e., the precautionary principle, the regulators have started several initiatives to shape the frameworks for future digital technologies, their applications, and the probably forthcoming new markets. In addition, the current regulations shaping the Single Market have to be adjusted and complemented by additional rules supporting the opportunities and prohibiting possible risks. The objective is not only to have an appropriate regulatory framework for the future of the Single Market of the European Union but also to influence the shaping of the regulations in other regions of the world following the logic of the Brussel effect, i.e., the attempt of globalize the regulatory frameworks generated by the European Union through the market mechanism. This objective has been partially achieved by the General Data Protection Regulation (GDPR) because some countries have already followed the principles of GDPR. Even companies have headquarters outside the European Union but are doing business in the Single Market adjusted their processing of privacy-related data. In addition to these regulatory principles, the European Union's regulatory framework is characterized by the unique collaboration between mandatory regulations and voluntary standards. Since its inception in the 1980s, the European New Legislative Framework (NLF) has played a crucial role in integrating the Single Market. The NLF sets a foundation that empowers stakeholders, particularly industry and other organizations, to participate actively in the standardization processes through the European Standardisation Organisations (ESOs). The introduction of Harmonised European Standards under Regulation 1025/2012 allows for a presumption of conformity with essential requirements, facilitating market access for a diverse range of products and mainly benefiting small and medium-sized enterprises (SMEs).

The digital landscape in Europe is undergoing a transformative phase as the European Union initiates a comprehensive revision of the conditions governing access to the European Union Single Market for software and hardware products. This edited volume is an important knowledge source within this evolution, providing stakeholders with a basic understanding of the evolving regulatory environment that shapes the market access to the Single Market in the context of digital technologies. Based on the current state of play, this book elaborates on the existing and forthcoming European Digital Regulations, differentiated into twelve areas including Cybersecurity, Data, and Artificial Intelligence. However, the dynamic nature of these regulations requires that stakeholders remain informed, as they may be revised, replaced, or even complemented by new regulatory initiatives. Each chapter can serve as a guide to complying with the current versions of the regulations and considering their implications for market access.

Particularly noteworthy is the analysis of the European Artificial Intelligence Act, which covers both High-Risk AI Systems and General Purpose AI Models. Via the published standardization request, this regulation allows stakeholders to influence its specification through active participation in standardization processes. Further requests related to other regulations are currently being prepared. Similarly to the Artificial Intelligence Act, the European Cyber Resilience Act and the European Radio Equipment Directive address the pressing need for robust cybersecurity measures and privacy standards, ensuring that products in the digital domain meet elevated health, safety, and environmental standards.

The book also covers topics such as Intelligent Transport Systems and the Networks and Information Systems Directive. These initiatives reflect the EU's aim to consider new technological advancements and address potential threats. The emphasis on cybersecurity strategy and the requirements set forth in this Directive underscore the importance of a cohesive and comprehensive approach to safeguarding the digital ecosystem.

Furthermore, the data-related regulations, including the Data Act and the General Data Protection Regulation (GDPR), show the potential critical role of data governance for the further development of digital economy. As data will have a pivotal role in driving innovation, competitiveness, and economic growth in the future, understanding the regulatory landscape related to data is essential for all stakeholders.

The insights presented in this volume are timely and essential for navigating the special characteristics of the European regulatory landscape. Various stakeholders, ranging from multinational enterprises to SMEs and research organizations, will find insights in the edited book that can support their decision-making and strategic planning in light of the evolving regulatory conditions relevant to accessing the European Union Single Market.

In conclusion, this edited book is an important resource for anyone seeking to understand the complexities of the digital regulations of the European Union. The efforts behind its creation by well-experienced industry experts reflect a commitment to advancing knowledge in an era characterized by rapid technological change and regulatory evolution. As the development of digital technologies will go on, as well as the accompanying regulatory landscape, all stakeholders must stay informed and engaged, ensuring that they not only comply with the new essential requirements but also actively contribute to shaping the future of the digital economy in Europe, e.g., via their involvement in standardization.

Prof. Dr. Knut Blind
Chair of Innovation Economics,
Faculty of Economics and Management
Technische Universität Berlin
Berlin, Germany

Preface

Conditions to access the European Union Single Market are currently undergoing substantial changes. A number of European Regulations are either in preparation or published in the European Union Official Journal, such that related *essential requirements* need to met after a transition period in order to maintain access to the European Union Single Market. New conditions related to Artificial Intelligence (through the European AI Act), Cybersecurity (through the activation of related articles in the European Radio Equipment Directive and the recently published Cyber Resilience Act) and others need to be understood by all stakeholders and related solutions supporting those *essential requirements* need to be integrated into products—otherwise access to the European Union Single Market will be lost.

Interestingly, available literature often focuses on the discussion of legal aspects related to those regulations. The present book takes a different approach. It is written by Engineers for Engineers, and other stakeholders who are interesting in a technical perspective. The objective is to understand technical implications of the new rules and to discuss implementation solutions which may help all stakeholders, including manufacturers and others, to fulfill the new requirements, while ensuring that civil servants can effectively monitor and enforce these regulations. This aims to bridge the gap between theory and practice, ensuring the effective implementation of digital regulations, even under resource constraints.

It is indeed expected that the understanding of those new regulations will further evolve over time, reflecting the dynamic nature of the regulatory landscape. European Standardisation Organisations, i.e. the European Telecommunications Standards Institute (ETSI), the European Committee for Electrotechnical Standardization (CENELEC) and the European Committee for Standardisation (CEN), are currently developing standards in support of the implementation of the new regulations according to standardisation requests issued by the European Commission. Related deliverables typically include so-called Harmonised European Norms—or Harmonised Standards—which provide presumption of conformity with related regulations once they are published in the European Official Journal. Those deliverables are key for all stakeholders, especially manufacturers, since they will provide detailed technical guidance on how to meet the *essential requirements* introduced by the regulations.

With new European Regulations directly relating to product related requirements, it becomes even more crucial for stakeholders to have a clear understanding of the underlying technical obligations. The present book will provide an introduction to the underlying framework and ecosystem, will explain how stakeholders can actively engage in shaping those deliverables and offer an overview on key European regulations including references to relevant standards deliverables and other related documents. It will thus support stakeholders to successfully navigate through the complex world of European Digital Regulations to maintain continued access to the European Union Single Market and contribute to a more operational regulatory landscape for the digital future.

Unterhaching, Germany — Markus Mueck
Paris, France — Christophe Gaie

Acknowledgments The editors would like to express their sincere gratitude to all authors and contributors who have generously shared their expertise and insights to enhance the quality of this book. Their collective efforts have resulted in a comprehensive and informative resource that reflects the diverse perspectives of various European countries. We are particularly grateful for their willingness to collaborate and contribute to a work that aims to promote a deeper understanding of European digital regulations.

Introduction

The European Commission, the European Parliament and the European Council are currently revising the conditions for software and/or hardware products and components to access the European Union Single Market, especially in the context of digital technologies. While market access conditions have been rather stable over past years for stakeholders, multiple changes are upcoming which need to be well understood—otherwise, products may lose access to the European Union Single Market. New *essential requirements*, ensuring products traded in the EU meet high health, safety, and environmental standards, are introduced in the fields of Cyber Security and Privacy (Cyber Resilience Act, Radio Equipment Directive Articles 3(3)(d/e/f), etc.), Artificial Intelligence (Artificial Intelligence Act, etc.), Data Management (Data Act, etc.) and others.

The present book provides an overview on existing and upcoming European Digital Regulations and summarizes key novel regulations to support all stakeholders to better understand the upcoming changes for accessing the European Union Single Market. Indeed, all stakeholders are concerned including multinational organizations, small and medium-sized enterprises (SMEs), research organizations, societal stakeholders, etc. These regulations are also crucial for government organizations and civil servants who will need to adapt their procurement processes, internal operations, and regulatory frameworks to comply with these new standards. To ensure perfect applicability, government services will need to adjust European regulations to the local context.

Chapter 1 provides an *Introduction to the European New Legislative Framework* (NLF). It is explained that the process started in the 1980s; the European Commission, under the leadership of President Jacques Delors, took the initiative to move forward in the European integration with a new project: the completion of the single market. Through a number of intermediate steps, the New Legislative Framework has eventually materialized which, among other benefits, allows stakeholders to actively shape the details of market access conditions through participation in related standardisation processes in the European Standardisation Organisations (ESOs). Through Regulation 1025/2012, ESOs are authorized to develop Harmonised European Norms—or Harmonised Standards—which provide presumption of conformity to the essential requirements of harmonised legislations. The process allows for self-declaration of conformity which proves to be highly beneficial for all stakeholders and especially supports small and medium-sized enterprises (SMEs).

Chapter 2 gives an *Overview on European Digital Regulations*, including existing Regulations and those currently in preparation. Those are clustered into twelve regulation categories as follows: (i) Research and Innovation, (ii) Industrial Policy, (iii) Connectivity, (iv) Data and Privacy, (v) Intellectual Property Rights (IPR), (vi) Cybersecurity, (vii) Law Enforcement, (viii) Trust and Safety, (ix) E-Commerce and Consumer Protection, (x) Competition, (xi) Media and (xii) Finance. Note that existing regulations may be revised or withdrawn at any time or complemented by new initiatives. The corresponding environment is highly dynamic.

Chapter 3 offers an *Introduction to the European Artificial Intelligence Act (EU AI Act)*, which was published in the European Union Official Journal in July 2024. The Act is thus rather recent, and its implementation was in the early stages at the time when the present book was written. It is important to understand that the EU AI Act specifically focuses on so-called High-Risk AI Systems and General Purpose AI (GPAI) Models of systemic risk. There will be ample opportunity for stakeholders to shape related access conditions to the European Union Single Market through participation in the corresponding standardisation processes in the European Standardisation Organisations according to Standardisation Requests issued by the European Commission. Furthermore, and in complement to the EU AI Act, an AI Liability Directive is under preparation which is expected to revise *current national liability rules*.

Chapter 4 provides an *Introduction to the European Cyber Resilience Act* (CRA) which was available as a draft at the time when the present book was written. The Cyber Resilience Act relates to *products with digital elements* and is likely to affect market access for (almost) every digital product and component, including hardware and software products or components. It is thus of utmost importance for all stakeholders to understand the new regulation framework, otherwise access to the European Union Single Market may be lost. With the objective to enforce a continued mitigation of newly identified vulnerabilities, the CRA introduces essential requirements to comply with at the time of placing products onto the market—and additionally requirements to be met over the lifetime of the product (no less than 5 years), including software updates for mitigating any newly discovered vulnerabilities and related reporting obligations.

Chapter 5 gives an *Introduction to European Radio Equipment Directive (RED)* and specifically focuses on newly activated Articles on Cybersecurity and Privacy for Radio Equipment, i.e. Radio Equipment Directive Articles 3(3)(d/e/f). Due to an obvious overlap with the new Cyber Resilience Act, it is likely that those RED requirements may be moved under the Cyber Resilience Act. Furthermore, the chapter details Radio Equipment Directive Article 3(3)(i) and Article 4 which relate to the combination of Hardware and Software; i.e. they address the case that software or firmware is loaded onto a product affecting the radio characteristics. Standards are summarized as they have been developed by the ETSI Technical Committee Reconfigurable Radio System (TC RRS); those are candidates to be transposed into

Harmonised European Norms—or Harmonised Standards—in case that the European Commissions decides to activate the related Articles through Delegated Acts.

Chapter 6 offers an *Introduction to European Intelligent Transport Systems Regulation*. The aim of the related initiative is to adapt to the emergence of new road mobility options, mobility apps and connected and automated mobility enabled through so-called Cooperative Intelligent Transport Systems (C-ITS). The objective is to stimulate the faster deployment of new, intelligent services, by proposing that certain crucial road, travel and traffic data is made available in digital format, such as speed limits, traffic circulation plans or roadworks. A specific focused is laid on Vehicle-to-Everything (V2X) communication systems and services. The related deployment was indeed slower than originally expected, mainly due to a technology battle between solutions based on IEEE 802.11p (a Wi-Fi derivate) and a direct-link based 3GPP solution (C-V2X).

Chapter 7 gives an *Introduction to the Networks and Information Systems 2 (NIS2) Directive*. The Directive mainly defines high level principles which need to be implemented on the level of European Union member states. At the time when the present book was written, the deadline for national transposition was imminent: According to NIS2 Article 41, by 17 October 2024, Member States shall adopt and publish the measures necessary to comply with the NIS2 Directive. The NIS2 Directive was implemented as a successor to the original NIS Directive with the objective to address the increasing complexity and severity of cyber threats. The original NIS Directive, while a step in the right direction, was deemed insufficient to address the evolving landscape, including (i) Rising Cyber Threats, (ii) Economic Impact, (iii) Harmonization, (iv) Expanded Scope, (v) Strengthened Enforcement, (vi) Incident Reporting and (vii) Supply Chain Security. A key component of the NIS2 Directive relates to the requirement for each Member State to implement a "National Cybersecurity Strategy". This National Cybersecurity Strategy needs to *provide for the strategic objectives, the resources required to achieve those objectives, and appropriate policy and regulatory measures, with a view to achieving and maintaining a high level of cybersecurity.*

Chapter 8 offers an *Introduction to European Data related Regulation*, spanning the Digital Markets Act, the Regulation on a Single Market for Digital Services, the Data Governance Act, the Data Act and the General Data Protection Regulation (GDPR). Especially the novel Data Act is of key importance to all stakeholders since it introduces broad essential requirements for managing data spaces across the European Union, including *(i) interoperability of data, of data sharing mechanisms and services, as well as of common European data spaces (Article 33), (ii) interoperability of data processing services (Article 35) and (iii) essential requirements regarding smart contracts for executing data sharing agreements (Article 36).*

Chapter 9 provides an *Introduction to the European Machinery Regulation*, as a successor to the earlier Machinery Directive. Similar to the far-reaching impact expected from the EU AI Act, the Machinery Regulation is anticipated to significantly influence machinery safety standards globally. As such, the new revision of Directive 2006/42/EC addresses new technologies including artificial intelligence, the Internet of things and robotics which raise new challenges in terms of product safety. Indeed, these new tools allow machinery products to apply real-time processing of information, problem solving, mobility, sensor systems, learning, adaptability, and capability of operating in unstructured environments.

Chapter 10 gives an *Introduction to the European Cybersecurity Act (CSA)*. The 2019 European Cybersecurity Act renews and redefines the mandate of the European Union Agency for Cybersecurity (ENISA) which was originally created in 2004 as the European Network and Information Security Agency. It aims to ensure a high level of cybersecurity, cyber resilience and trust in the European Union, specifically through voluntary European cybersecurity certification schemes for information and communications technology (ICT) products, services and processes. Multiple Regulations indeed rely on such CSA based Certification Schemes for compliance to essential requirements. The existing or developing Certifications Schemes are summarized: (i) The European Common Criteria-based cybersecurity certification scheme (EUCC), (ii) EU Cloud Services Scheme (EUCS) and (iii) 5G Cybersecurity Certification (EU5G Scheme). Furthermore, an outlook on planned future Certification Schemes is given.

The upper chapters have the objective to bring clarity into the complex and interdependent Digital Regulations in Europe. It is essential for all stakeholders to have a good level of understanding of the underlying principles and new essential requirements, defining the access to the European Union Single Market. Continued compliance to the essential requirements is of utmost importance to maintain access to the European Union Single Market.

The research project VERGE (Grant Agreement 101096034—Smart Networks and Services Joint Undertaking under the European Union's Horizon Europe research and innovation program) has partially supported the writing of this book, which addresses very relevant aspects tackling the AI-powered evolution towards beyond 5G open and secure edge architectures.

Thank you for choosing our book.
We hope you will enjoy the read!

Markus Mueck
Christophe Gaic

Contents

Editors and Contributors

About the Editors

Dr. Markus Mueck received the Dipl.-Ing. and ing. dipl. degrees from the University of Stuttgart, Germany, and the Ecole Nationale Supérieure des Télécommunications (ENST), Paris, France, respectively in 1999. In 2006, he received the Doctorate degree of ENST in Communications. From 1999 to 2008, Dr. Mueck was Senior Staff member and Technical Manager at Motorola Labs, Paris, France. In this role, he contributed actively to various standardization bodies, namely Digital Radio Mondiale, IEEE 802.11n, etc., and led the creation of the novel standardization group IEEE P1900.4 in the area of Cognitive Radio and Software Defined Radio (SDR). He also contributed to numerous European Research projects, namely as Technical Manager of IST-E^{2}R II (19 MEuros budget) and as overall technical leader for the definition of IST-E^3 (20 MEuros budget). In 2008, Dr. Mueck joined Infineon Technologies, Munich, Germany, where he was head of delegation for IEEE 802.16m and transitioned to Intel Mobile Communications in 2011 in the framework of an acquisition by Intel. Currently, Dr. Mueck is Director and Principal Engineer of Intel Laboratories, he acts as Chair of the European Telecommunications Standards Institute (ETSI) Board, and he is Chair of the ETSI Operational Co-ordination Group Artificial Intelligence (OCG AI) coordinating the overall effort of the organization on AI, member of ETSI's Delegation to 3GPP PCG/OP and Adj. Professor of University of Technology, Sydney, Australia. Dr. Mueck was leading the effort on Artificial Intelligence and Machine Learning of the European-funded research Flagship Project Hexa-X preparing the design of 6th Generation Mobile Networks (6G).

Dr. Christophe Gaie earned his Ph.D. in Telecommunications from Université Paris Sud—Saclay in 2010, following research collaboration between Motorola and Supélec. He also holds a double degree in Engineering and a Master of Research from INP Bordeaux—Université Bordeaux, obtained in 2006. Currently, he serves as the Head of the Engineering and Digital Innovation Division at the Prime Minister's

Office. Previously, he led the digital applications office at the Centre interministériel de services informatiques relatifs aux ressources humaines (CISIRH) and managed the interministerial exchange platform at the Direction Interministérielle du Numérique (DINUM). With over a decade of experience at the Direction Générale des Finances Publiques (DGFiP), he initially worked as an IT architect and later headed the Income Tax and Withholding Tax team. His main research interests include e-Government, modernization of Information and Telecommunication Systems, API development and usage, Artificial Intelligence, and fraud detection.

Contributors

Ahmad Al Shamsi Telecommunications and Digital Government Regulatory Authority (TDRA), Abu Dhabi, United Arab Emirates

Sultan Albalooshi Telecommunications and Digital Government Regulatory Authority (TDRA), Abu Dhabi, United Arab Emirates

Shaikha Almulla Telecommunications and Digital Government Regulatory Authority (TDRA), Abu Dhabi, United Arab Emirates

Mohammad Alshamsi Telecommunications and Digital Government Regulatory Authority (TDRA), Abu Dhabi, United Arab Emirates

Lina Bariah Khalifa University, Abu Dhabi, United Arab Emirates

Merouane Debbah Khalifa University, Abu Dhabi, United Arab Emirates

Stephane Du Boispean Intel Corporation SA/NV, Kontich, Belgium

Stéphane Du Boispéan Intel Corporation SA/NV, Kontich, Belgium

Stéphane du Boispéan Intel Corporation SA/NV, Brussels, Belgium

Dimitris C. Gkikas University of Patras, Patras, Greece

Christophe Gaie Head of the Engineering and Digital Innovation Division, Prime Minister Services, Paris, France

Christophe Gossard John Deere, Mannheim, Germany

Markus Mueck Intel Deutschland GmbH, Munich, Neubiberg, Unterhaching, Germany

Taylor Roberts Intel UK, Swindon, UK

John Roman Intel Corp., Santa Clara, USA

Acronyms

3GPP	3rd Generation Partnership Project
5GAA	5G Automotive Association
AADHAAR	A twelve-digit unique identity number
AI	Artificial Intelligence
ASIC	Application-Specific Integrated Circuits
ATEX	ATmosphère EXplosible (explosive atmospheres)
BSI	Bundesamt für Sicherheit in der Informationstechnik (Federal Office for Information Security, Germany)
BTP	Basic Transport Protocol
CA	Certificate Authority
CAM	Cooperative Awareness Message
CAP	Conformity Assessment Procedure
CCAM	Connected, Cooperative and Automated Mobility
CE	Conformité Européene (European Conformity)
CEN	European Committee for Standardization
CENELEC	European Committee for Electrotechnical Standardization
CEPT	European Conference of Postal and Telecommunications Administrations
CFR	Code of Federal regulations
C-ITS	Cooperative Intelligent Transport Systems
CNC	Computerised Numeric Controllers for machine tools
COD	Ordinary legislative procedure
COVID	Coronavirus disease
CRA	Cyber Resilience Act
CRD	Directive on Consumer Rights
CSA	Cyber Security Act
CSAM	Child Sexual Abuse Material
CSIRT	Computer Security Incident Response Team
CSL	Communication Service Layer
DCC	Decentralized Congestion Control
DCS	Distributed Control Systems

DENM	Decentralised Environmental Notification Message
DEP	Digital Europe Programme
DMA	Digital Market Act
DSA	Digital Services Act
DSRC	Dedicated Short-Range Communications
EC	European Commission
EC3	European Cybercrime Centre
ECC	Electronic Communications Committee
EDIB	European Data Innovation Board
EE	Essential Entity
EEA	European Economic Area
EEC	European Economic Community
EG	ETSI Guide/Expert Group
EHDS	European Health Data Space
EHSR	Essential Health and Safety Requirement
eIDAS	electronic IDentification, Authentication and trust Services
EMC	Electromagnetic Compatibility
EN	European Norm
ENISA	European Union Agency for Cybersecurity
ESO	European Standards Organisation
ETSI	European Telecommunications Standards Institute
EU	European Union
EU5G	5G Cybersecurity Certification
EUCC	European Common Criteria-based cybersecurity certification
EUCS	EU Cloud Services Scheme
EU-CyCLONe	European Cyber Crisis Liaison Organisation Network
eUICC	Embedded Universal Integrated Circuit Card
EUOJ	European Union Official Journal
FCC	Federal Communications Commission
FPGA	Field-Programmable Gate Arrays
GCC	Gulf Cooperation Council
GDPR	General Data Protection Regulation
gMURI	generalized MUlti-Radio Interface
GPAI	General Purpose Artificial Intelligence
GPS	Global Positioning System
GSMA	Global System for Mobile Communications Association
HEN	Harmonised European Norm
HS	Harmonised Standard
HSM	Hardware Security Module
IACS	Industrial Automation and Control Systems
ICT	Information and Communications Technology
IDC	International Data Corporation
IE	Important Entity
IEC	International Electrotechnical Commission
IEEE	Institute of Electrical and Electronics Engineers

IoT	Internet of Things
IP	Internet Protocol
IPC	Internal Production Control
IPR	Intellectual Property Rights
IPv6	Internet Protocol version 6
ISO	International Organization for Standardization
IT	Information Technology
ITRE	Industry, Research and Energy
ITS	Intelligent Transport Systems
ITU	International Telecommunication Union
JTC	Joint Technical Committee
KDB	Knowledge Database
MAC	Medium Access Control (Layer)
MEITY	Ministry of Electronics and Information Technology (India)
METI	Ministry of Economy, Trade and Industry (Japan)
MIC	Ministry of Internal Affairs and Communications (Japan)
MiCA	Crypto-assets Regulation
ML	Machine Learning
MoST	Ministry of Science and Technology (China)
NANDO	New Approach Notified and Designated Organisations
NCC	National Coordination Centres
NESAS	Network Equipment Security Assurance Scheme
NET	European telecommunication norms
NIS	Networks and Information Systems
NIST	National Institute of Standards and Technology
NLF	New Legislative Framework
NOI	Notice of Inquiry
NPRM	Notice of Proposed Rule Making
NR	New Radio
NR-V2X	New Radio-V2X
OBU	On-Board Unit
OEM	Original Equipment Manufacturer
OJEU	Official Journal of the European Union
OS	Open Source
PAG	Pre Approval Guidance
PC5	Direct communication technology based on mobile network (3GPP)
PDCP	Packet Data Convergence Protocol
PHY	Physical Layer
PKI	Public Key Infrastructure
PLC	Programmable Logic Controllers
PLD	Product Liability Directive
PP	Protection Profile
PSD	Payment Service Directive
RAP	Radio Application Package

RE	Radio Equipment
RED	Radio Equipment Directive
RF	Radio Frequency
RLAN	Radio Local Area Networks
RLC	Radio Link Control
RRS	Reconfigurable Radio Systems
RSPP	Radio Spectrum Policy Programme
RSU	Road Side Unit
SAE	Society of Automotive Engineers
SAI	Securing AI (ETSI Technical Committee)
SAS	Security Assurance Scheme
SCADA	Supervisory Control and data Acquisition systems
SCMS	Security Credential Management System
SDL	Secure Development Lifecycle
SDO	Standards Development Organization
SIEM	Security Information and Event Management
SIM	Subscriber Identification Module
SMCS	Single Market Compliance Space
SME	Small and Medium-sized Enterprises
SMEI	Single Market Emergency Instrument
SoA	State of the Art
SPaT	Signal Phase and Timing
SR	Standardisation Request
SRD	Short-Range Devices
SRTI	Safety-Related Traffic Information
SSTP	Safe and Secure Truck Parking
STEP	Strategic Technologies for Europe Platform
TC	Technical Committee
TCB	Telecommunication Certification Bodies
TCP	Transmission Control Protocol
TDRA	Telecommunications and Digital Government Regulatory Authority, UAE
TG	Thematic Group
UAE	United Arab Emirates
UCPD	Unfair Commercial Practices Directive
UDP	User Datagram Protocol
UICC	Universal Integrated Circuit Card
UK	United Kingdom
UMTS	Universal Mobile Telecommunications System
US	United States
Uu	UMTS air interface
UWB	Ultra-Wide Band
V2I	Vehicle-to-Infrastructure
V2N	Vehicle-to-Network
V2P	Vehicle-to-Person

V2V	Vehicle-to-Vehicle
V2X	Vehicle-to-Everything
VPN	Virtual Private Network
WAVE	Wireless Access in Vehicular Environments
Wi-Fi	Wireless Fidelity

Chapter 1
Introduction to the European New Legislative Framework

Stéphane du Boispéan, Markus Mueck, and Christophe Gaie

Abstract The chapter provides an overview on the New Legislative Framework (NLF), the history and context that led to the adoption of this legal and "soft-law" framework, and the rationale behind the decisions taken by the legislator. The NLF indeed creates a unique framework for the implementation of legislative projects on the level of the European Union. While other regions rely on a strict type approval process for market access, the European framework ensure a close interaction between policy makers and stakeholders. As instantiated through European Regulation 1025/2012, stakeholders are able to engage into an open and transparent process of developing so-called Harmonized European Norms—or Harmonized Standards—which provide presumption of conformity with a related Legislation once they are listed in the European Union Official Journal. Instead of relying on external services, typically provided by accredited Notified Bodies, self-declaration of conformity is possible which provides a straightforward process to market access that is especially well suited to Small and Midsize Enterprises.

Keywords Regulation · Market access · European Union single market · CE marking · Conformity · Safety

---

S. du Boispéan (✉)
Intel Corporation SA/NV, Brussels, Belgium
e-mail: Stephane.du.Boispean@intel.com

M. Mueck
Intel Deutschland GmbH, Neubiberg, Germany
e-mail: Markus.Dominik.Mueck@intel.com

C. Gaie
Head of the Engineering and Digital Innovation Division, Prime Minister Services, Paris, France

M. Mueck and C. Gaie (eds.), *European Digital Regulations*, Intelligent Systems Reference Library 265, https://doi.org/10.1007/978-3-031-80809-8_1

1.1 Introduction

In the 1980s, the European Commission, under the leadership of President Jacques Delors, took the initiative to move forward in the European integration with a new project: the completion of the single market. Such a project was endorsed by the Head of States and Governments and made a consensus between the countries willing to keep the European project merely an economic one based on free trade, and those willing to go further in the political integration. The latter considered the single market as a step towards a stronger level of decision-making at European level, while the first group could hardly disagree with the idea of an enhanced free trade zone. For several decades, the single market has been considered as one of the biggest political achievements of the European integration [1, 2].

The main idea behind the single market is the freedom of movement within the European Union (back then European communities) for goods, services, persons, and capitals. The single market was achieved through various legal acts and political initiatives. This chapter will have a closer look at the political and regulatory framework behind the free movement of goods in the EU. It will analyse the principles behind the NLF, from their origins to the most recent developments.

1.2 The New Approach

In order to achieve the free movement of goods within the single market, the European Council, representing the Heads of States and Governments, adopted in May 1985 a resolution "*on a new approach to technical harmonization and standards*" [3] commonly since known as the New Approach. This resolution makes the link between standards and the free movement of goods, as it explicitly quotes technical barriers to trade as an issue to be solved. It also states that the New Approach "*provides for reference to standards (...) for the purpose of defining the technical characteristics of a product*". Furthermore, it requests the Commission to set up a policy framework related to the conformity of products. The May 1985 resolution on the New Approach could therefore be considered as a starting point for the work that would lead years after to the creation of the New Legislative Framework. It is of highest importance for standards negotiations [4].

The core idea behind the New Approach, as opposed to what is often referred to as the "*Old approach*" [5], is to separate the high-level compliance requirements for products (so-called essential requirements) from the technical standards that define in detail how to comply with the regulatory framework. It should be noted that priority is given to standards at European level, while not ruling out national standards ("*preferably European but if necessary national*"). Thus, the New Approach is illustrated in Fig. 1.1.

The Annex II of the May 1985 Council resolution, titled "*Guidelines for a new approach*" clearly states that "*legislative harmonization is limited to the adoption, by*

Fig. 1.1 Separation of high-level compliance requirements from the technical standards (new approach)

means of Directives (…) of the essential safety requirements (or other requirements in the general interest) with which products put on the market must conform, and which should therefore enjoy free movement throughout the Community" and that on the other hand "*the task of drawing up the technical specification needed for the production and placing on the market of products conforming to the essential requirements established by the Directives, while taking into account the current stage of technology, is entrusted to organizations competent in the standardization area*".

The resolution also lists the main elements which should make the content of future Directives harmonizing the essential requirements for products. The most relevant of them are illustrated in Fig. 1.2:

Key principles of the May 1985 Council resolution are further outlined below:

- A definition of the scope of each harmonized legislation, stating which products are covered and which hazards should be prevented. The Council sets also the objective of having a limited number of *lex specialis*. At the same time, the resolution could also be read as keeping the door open to several acts covering the same products but for different hazards.
- Rules for placing on the market of products: the Council mentions the idea of "spot checks" carried out by the Member States, which correspond to what we know today as market surveillance. On the other hand, it introduces a high-level and generic principle that the placing on the market should in theory not be preceded by controls (with some exceptions possible). This means a separation of roles is established between the operators placing the product on the market and the national authorities, who are first and foremost in charge of carrying out *ex post* controls.
- Essential safety requirements and free movement of goods: In line with the idea of separating essential requirements from technical specifications, the future Directives should, according to the Council define clear requirements that need to be

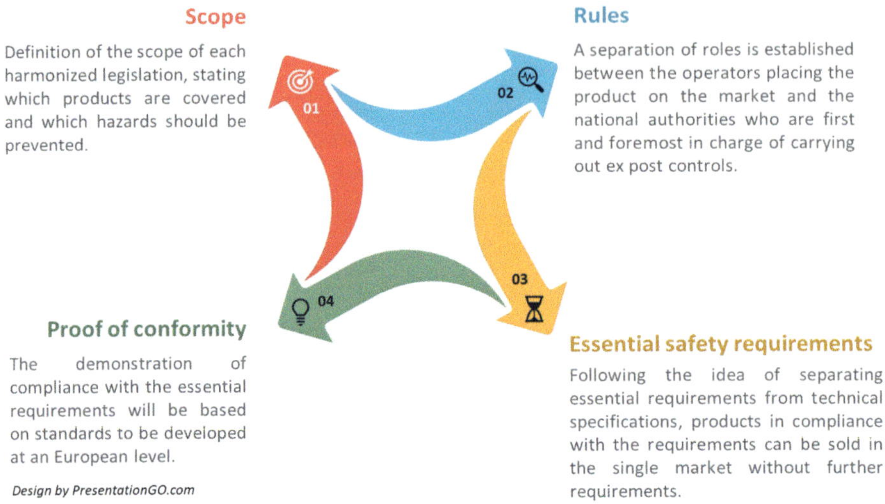

Fig. 1.2 Basic principles of the May 1985 Council resolution

precise. The logical consequence is that products in compliance with the requirements can be sold in the single market without further requirements, except in some limited cases. The Member States will see their ability to block the movement of products heavily restricted. This is a part of the New Approach that should not be underestimated: the main objective of the Directives to be adopted under the New Approach is the free movement of products and the possibility for the manufacturers to sell their goods beyond the existing national markets.

- Means of proof of conformity through standards: The Member States accept the idea that in order to demonstrate compliance with the essential requirements, standards will have to be developed at European level and that national standards play a role only in limited cases, for instance where there are no European standards. They also accept the idea that not all national standards will be applicable and enforced. Instead, the first outline of a procedure in order to mutually accept some national standards as valid is made.

The Council also makes a list of sectorial priorities in which legislative acts should be adopted. Those are for instance "*engineering products and building materials*". It should be noted that the future legislations are meant to target families of products rather than singular goods, in order to create a simplified framework and harmonization. This is also understood as a difference with the Old Approach and its patchwork of product-based rule [6].

1.3 From the New Approach to the New Legislative Framework: Example of Radio Equipment

Following the adoption of the Council resolution on the New Approach and on standards, the Commission, as EU body entrusted by the Treaties with the right to table legislative initiatives, started its work and tabled several Directives harmonizing the requirements for specific products. Examples include the Electromagnetic Compatibility Directive 89/336/EEC [7], the Telecommunication Equipment Directive 86/361/EEC [8] or the Directive on active implantable medical devices 90/385/EEC [9]. While these legislations have been amended over time (several times for the majority of them), the first series of legislations adopted are rather short texts setting generic principles in line with the New Approach resolution. It took several years to slowly build a series of legislations that implemented the high-level principles of the New Approach.

Directives and Regulations are the two most commons legal instruments of EU law under the EU Treaties. The main difference is that a Directive is a legal act that needs to be transposed into national law before producing legal effects (with some exceptions), while a Regulation is directly applicable in all Member States. Traditionally, Directives, especially in the 1980's, were rather short texts leaving flexibility to the Member States. However, in the last years, the length and precision of the Directives have increased. At the same time, several important EU Regulations (such as for instance the GDPR) are enforced primarily at national level by authorities that are under the control of the Member States. This could in some cases mean some diverging implementation of the same legal text. The difference between Directives and Regulations, while still being valid, tends to be less stringent than in the past.

Directive 86/361/EEC [8] on the telecommunication Equipment is a good example of this process. Officially dealing with the "*mutual recognition of type approval*" of related equipment, the original Directive sets up a series of definitions (Article 2) which allow to clarify the scope and defines in its Article 2, Paragraph 17, the essential requirements, limited (in the spirit of the 1985 resolution) to high-level requirements: user safety, protection of employees of public network operators, protection of network from harm and interworking of equipment. The Commission is tasked by the Directive to identify international standards to be harmonized and to request standard setting bodies to develop European norms, called at that time "*common conformity specifications*" (Article 6). European telecommunication norms (NET) adopted by the standard setting bodies such as the CEPT are considered valid common conformity specifications and fulfil the essential requirements. The Directive logically prevents Member States from imposing additional conformity assessment mechanisms for products that have been assessed in accordance with the common conformity specification (Article 6). Such a principle comes with exceptions that are clearly defined in the Article 8 of the Directives. The legislator sets up a procedure for exchange of information and discussion between the Member States and the Commission.

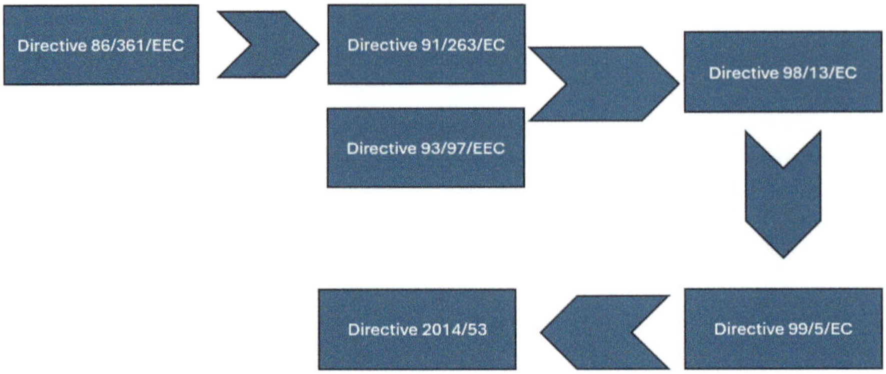

Fig. 1.3 Evolution of the EU regulation until the adoption of Directive 2014/53

The Directive has been amended several times since then, with various revisions and repeal. It was amended first in 1991, then merged with the 1993 Directive on satellites earth station equipment and then revised in 1998, 1999 and 2024. The process ultimately led to the adoption of the most recent text: Directive 2014/53 [10], the so-called Radio Equipment Directive (RED). This evolution is illustrated in Fig. 1.3.

The new Directive 91/263/EC [11] added new essential requirements such as electromagnetic compatibility and provisions on the use of spectrum, as well as a revised wording around the presumption of conformity for products in compliance with standards harmonized at European level. It also sets some principles for the recognition and work of conformity assessment bodies. Slowly, the rationale behind the legal text moves from the prohibition of national standards perceived as barriers to trade, to the establishment of a European system of standards, where national standards still play a certain role to support conformity assessment at a European level instead of a national one.

Directive 98/13/EC [12] merges Directive 91/263/EC [11] with Directive 93/97/EEC [13] and aims at bringing "*clarity and rationality*" in the legal framework by codifying the two texts in one. It brings less changes to the legal framework. Directive 99/5/EC [14], which repeals Directive 98/13/EC [12], adapts (according to its wording) the legal framework to the new reality of the market, driven by technological innovation. For instance, it mentions as essential requirements, among other things, the access to emergency services and the protection against fraud. The Directive also mentions considerations around innovation and essential requirements, stating the objective to find a balance between the two objectives: on the one hand "*not to inhibit technological innovation or the meeting of the needs of a free-market economy*" (Recital 13), on the other hand that "*radio equipment and telecommunications terminal equipment should not represent an avoidable hazard to health*" (Recital 14). It also introduces a structured framework regarding the conformity assessment of radio equipment (Article 10) with different modules in place, chosen by the manufacturer (Article 10).

Over the years, European Directives in general have become longer and more detailed, with requirements being laid out more precisely, what could be interpreted also in a more prescriptive way). Directive 91/263/EEC [11] and then Directive 99/5/EC [14] on *radio equipment and telecommunications terminal and the mutual recognition of their conformity* make no exception to this and is more detailed than the original Directive 86/361/EEC.

Directive 2014/53 [10], the so-called Radio Equipment Directive (RED), was adopted together with several other pieces of legislation related to harmonized products (electromagnetic compatibility, lifts, ATEX[1] low voltage, pressure equipment). It constitutes the current EU legal framework for the placing on the market of radio equipment. The Directive follows a structure which is similar to other legislations, in line with the 2008 Decision (see X.4). It starts by defining the scope (Article 1), then the key concepts (Article 2) and continues with the essential requirements in Article 3. The legislator, like in the previous texts, has adapted the requirements to the technological innovation and includes the absence of harm to the network, the protection of personal data and the avoidance of fraud (Article 3. 3 d, e, f). New provisions are included in the Directive, such as detailed obligations for manufacturers, their representatives and/or importers (Article 10–12). The Directive dedicates a full chapter (III) to the conformity assessment of radio equipment, which lists the different possibilities for the manufacturer to demonstrate compliance with the essential requirements (Article 17). It also lists the detailed requirements that apply to conformity assessment bodies designated under the Directive (Notified Bodies). The present book contains a specific chapter providing a deep-dive into the Radio Equipment Directive where further information is available.

The example of radio equipment shows how the priorities of the legislator evolved over time while keeping several basic principles. It started with a legislation aiming at ensuring that Member States would not prevent the free movement of goods from one Member State to another, especially by restricting the use of national standards. It slowly evolved to a legal framework where standards and norms adopted or referenced at European level play a key role and are considered as enablers. The competency for setting rules regarding the placing on the market of radio equipment has evolved from a national to a European one. This is a concrete example of the completion of the internal market through legislative action.

1.4 The 2008 New Legislative Framework (NLF)

Once the principle of the harmonization through legal acts was established in the Resolution of 1985, several products, beyond the example of radio equipment, were harmonized over the years. In order to facilitate the adoption of future legislation, the Council adopted over the years several Decisions defining procedures for the

[1] ATEX is an abbreviation from the French, ATmosphère EXplosible or explosive atmospheres, translated into English.

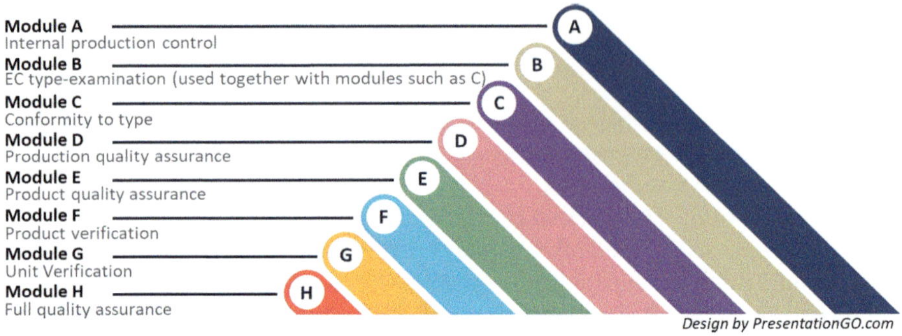

Fig. 1.4 Modules laying down in detail the procedure for the assessment of the conformity of a product against essential requirements

verification of the compliance of products with the essential requirements. The latest version constitutes what is commonly referred to as the NLF, or New Legislative Framework of 2008 [15].

Decision 90/683/EEC [16] is to be understood in the political context of the time. It starts with referring to the deadline of the completion of the internal market by the end of 1992. Furthermore, the Decision is presented by the Council as a text helping at ensuring that consumers are protected and that manufacturers can take advantages of the internal market. The use of European conformity assessment mechanisms, especially through the so-called "*technical harmonization Directives*", is presented by the Council as an important instrument of the internal market. The Decision is made mostly of annexes detailing so-called Modules laying down in detail the procedure for the assessment of the conformity of a product against essential requirements of a harmonized Directive.

The modules are named by alphabetical order and illustrated in Fig. 1.4:

Decision 93/465/EEC [17] updates the 1990 Decision. It details in particular requirements related to the CE marking. This marking is to be affixed by the operator in charge of the conformity assessment (manufacturer, its representative, or in some cases the Notified Body) on the product in order to inform the consumers that the product is following the harmonized legislation.

In 2008, the legislator introduced two new legal instruments repealing the previous decisions, with the objective to improve the market surveillance (activities of the Member States ensuring ex-post compliance checks), to ensure a harmonized framework for the accreditation of conformity assessment bodies, to reform the rules related to the use of the CE marking, and to introduce templates for the provisions related to conformity assessment of future harmonized legislations. The 2008 New Legislative Framework (NLF) is made of two original texts: Decision 768/2008 [18] which repeals Decision 93/465/EEC [17], and Regulation 765/2008 [19] on accreditation and market surveillance which repeals the older Regulation 339/93 [20].

Decision 768/2008 [18] aims at expanding the principles of the New Approach to additional fields and products. Unlike the previous legal or non-binding texts adopted in the previous decades, it is less focussing on establishing the internal market (by removing national rules that would be barriers to trade) but more on consolidating it and establishing mechanisms for the placing on the market of products at the European level. The Decision is defined as "*a general framework of a horizontal nature for future legislation harmonizing the conditions for the marketing of products and a reference text for existing legislation.*" Its main objective is to ensure consistencies between the legislations harmonizing requirements for products, especially in the field of conformity assessment. The Decision repeats the high-level principles of the single market for products. For instance, it states that harmonized legislation should "*limit itself to the expression of essential requirements*" (Recital 8) and use harmonized standards "*more detailed technical specifications*". The separation between essential requirements and technical specifications remains therefore at the core of the product safety framework.

The different modules that can be used for conformity assessment are updated in the Annex of the Decision and explicitly referenced as future provisions of harmonized legislations (Article 4). The legislator, when making future legislations, will have to choose modules from the Annex II of the 2008 Decision. In particular, the degree of risk associated with the product will have to be considered in order to limit the burden on the manufacturer.

This balancing between the objective of protecting consumers and helping manufacturers is not new and has been mentioned in previous acts, but it is still present.

The Annexes of the Decision provide a series of templates with provisions that can easily be copied in future harmonized legislations. Annex I in particular ("*Reference provisions for community harmonization legislation for products*") is written like a future Directive with detailed structure that includes several chapters as illustrated in Fig. 1.5:

Further details on the upper templates are discussed below:

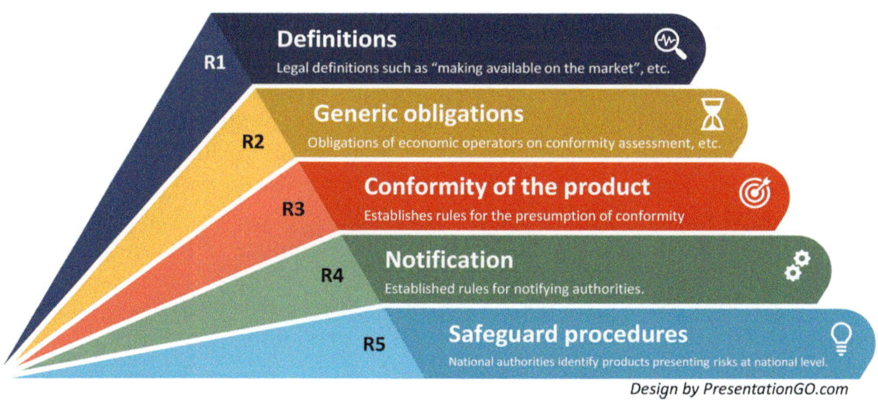

Fig. 1.5 Templates provided by Decision 768/2008 for future harmonized legislations

- Chapter R1 on Definitions includes legal definitions of several concepts used in the context of product safety legislation, such as "*making available on the market*", "*placing on the market*", "*manufacturer*", "*conformity assessment*", "*CE marking*". By doing so, the NLF Decision aims at ensuring that all future harmonized legislation will use the same legal definitions.
- Chapter R2 establishes the generic obligations of economic operators (manufacturers, authorised representatives, importers, distributors) regarding mostly the conformity assessment and the responsibility towards national surveillance authorities.
- Chapter R3 ("*Conformity of the product*") establishes rules for the presumption of conformity of a product and the use of standards. It also sets principles for the CE marking and the declaration of conformity, which are regulated in form of Annexes of the Decision.
- Chapter R4 ("*Notification of conformity assessment bodies*") establishes the rules regarding the designation of the national authorities overseeing the conformity assessment bodies (notifying authorities), the internal requirements of the conformity assessment bodies designated under the specific legislation (notified bodies) including the independence, the requirements related to the personnel and the impartiality (including the subcontracting of tasks). The chapter also set the rules for the designation of the notified bodies and the establishment of a database. This database originally named New Approach Notified and Designated Organisations (NANDO) still exists under the name Single Market Compliance Space (SMCS) [21].

Chapter R5 ("Safeguard procedures") sets future rules for the cases where national authorities identify products presenting risks at national level, with a procedure in place to ensure that action is taken by the relevant actor and that the information is exchanged at European level. The procedure also opens the door for Member States to challenge the decision of other authorities. This mechanism also limits the powers of national authorities, in line with the rational behind the establishment of the internal market initiated several decades before.

Annex II of Decision 768/2008 [18] lists the different modules that can be picked by the legislator in future harmonized legislations. The modules apply to both the design and the production phase of the products. While Module A provides a self-assessment through the manufacturer, the others generally include the intervention of a notified body (nothing prevents the manufacturer to use the services of a notified body even in Module A).

Annex III of Decision 768/2008 [18] lists the template of the declaration of conformity. This means that all declaration of conformity for all future harmonized product safety legislations should have a similar structure and the similar information (Table 1.1).

Decision 768/2008 [18] sets a template for future legislation. The legislator is in theory free to deviate from such a template but must justify this (Recital 6). This makes it harder for future legislations to depart from the common provisions of the Decision.

Table 1.1 Template of the declaration of conformity a defined in Annex III of Decision 768/2008 [18]

Annex III: EC Declaration of Conformity

1. No … (unique identification of the product):
2. Name and address of the manufacturer or his authorised representative:
3. This declaration of conformity is issued under the sole responsibility of the manufacturer (or installer):
4. Object of the declaration (identification of product allowing traceability. It may include a photograph, where appropriate):
5. The object of the declaration described above is in conformity with the relevant Community harmonization legislation: …
6. References to the relevant harmonized standards used or references to the specifications in relation to which conformity is declared:
7. Where applicable, the notified body … (name, number) … performed … (description of intervention) … and issued the certificate: …
8. Additional information:

Signed for and on behalf of:…………………
(place and date of issue):
(name, function) (signature):

Since 2008, 27 legislations have been either written or updated with new texts in order to be adapted to the New Legislative Framework. The European Commission keeps a list of the acts that are, according to the Commission, "*aligned with, or based on*" the provisions of the Decision.

Those are [22]:

- Toy Safety—Directive 2009/48/EU;
- Transportable pressure equipment—Directive 2010/35/EU;
- Restriction of Hazardous Substances in Electrical and Electronic Equipment—Directive 2011/65/EU;
- Construction products—Regulation (EU) No 305/2011;
- Pyrotechnic Articles—Directive 2013/29/EU;
- Recreational craft and personal watercraft—Directive 2013/53/EU;
- Civil Explosives—Directive 2014/28/EU;
- Simple Pressure Vessels—Directive 2014/29/EU;
- Electromagnetic Compatibility—Directive 2014/30/EU;
- Non-automatic Weighing Instruments—Directive 2014/31/EU;
- Measuring Instruments—Directive 2014/32/EU;
- Lifts—Directive 2014/33/EU;
- ATEX—Directive 2014/34/EU;
- Radio equipment—Directive 2014/53/EU;
- Low Voltage—Directive 2014/35/EU;
- Pressure equipment—Directive 2014/68/EU;
- Marine Equipment—Directive 2014/90/EU;
- Cableway installations—Regulation (EU) 2016/424;
- Personal protective equipment—Regulation (EU) 2016/425;
- Gas appliances—Regulation (EU) 2016/426;

- Medical devices—Regulation (EU) 2017/745;
- In vitro diagnostic medical devices—Regulation (EU) 2017/746;
- EU fertilising products—Regulation (EU) 2019/1009;
- Drones—Commission Delegated Regulation (EU) 2019/945 on unmanned aircraft systems and on third-country operators of unmanned aircraft systems;
- Batteries—Regulation (EU) 2023/1542;
- Machinery—Regulation (EU) 2023/1230;
- Ecodesign requirements for sustainable products—Regulation (EU) 2024/1781.

Other legislations could be added in the future.

The NLF framework officially consists also of Regulation 765/2008 [19] on the accreditation of Notified Bodies and market surveillance. This Regulation, which repeals the older Regulation 339/93 [20], harmonizes the procedure for the accreditation of conformity assessment bodies by setting common rules related to the functioning of the national authorities tasked with authorizing the conformity assessment bodies to operate: the accreditation bodies. Accreditation bodies must be independent, not operate as a conformity assessment body, not compete among each others, and are subject to a system of peer review (evaluation by a European-wide body coordinating the accreditation bodies at national level).

The roles and responsibilities of the accreditation bodies compared to the conformity assessment bodies (also valid for notifying authorities and notified bodies) is summarised as follow by the Dutch accreditation body [23] and illustrated in Fig. 1.6:

Chapter III of the Regulation established provisions that will apply in any case to all harmonized legislations (unless the legislation establishes specific provisions, but it seems that the intention of the legislator is to set the same rules for all products in order to ensure coherence). This chapter establishes a coordination between the Member States regarding market surveillance and the exchange of information between the authorities. It also (Article 21), in line with the provisions of previous legislations, limits the powers of national market surveillance authorities by requesting that restrictive measures are proportionate and allow the economic operator to be consulted. The Regulation also empowers market surveillance authorities to take actions regarding products imported into the EU (named Community in 2008, until the adoption of the Treaty of Lisbon in 2009), especially by prohibiting the release for free circulation.

Regulation 765/2008 [19] was amended by Regulation 2019/1020 [17] on market surveillance and compliance of products. The two texts continue to exist in parallel. Regulation 2019/1020 [17] is defined as a *lex specialis* (Recital 4), which means it will apply to all products that do not have similar market surveillance requirements in their harmonized legislation. The Regulation puts measures in place especially for products imported from non-EU countries through online sales (Fig. 1.7).

Article 16 lists the different measures that can be taken by the authorities, including a prohibition of sales or the physical destruction of the products. Articles 25 to 28 of the Regulation lists the powers of the market surveillance regarding products entering the EU internal market. These provisions are rather similar to the provisions of Regulation 765/2008 [19].

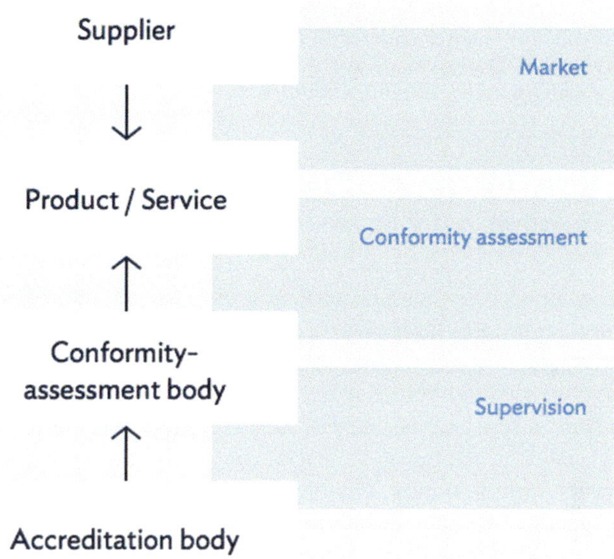

Fig. 1.6 Roles and responsibilities of the accreditation bodies

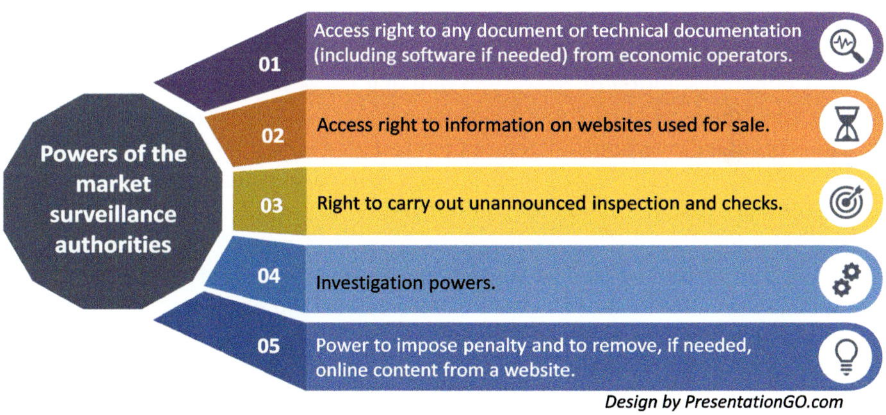

Fig. 1.7 Powers of the market surveillance authorities

Following the adoption of the NLF in 2008 and the adaptation of the existing harmonized legislations to the templates that were created, in particular by Decision 768/2008 [18], the European Commission has initiated between 2020 and 2022 an evaluation of the 2008 framework, with a potential revision as a change. The current texts could therefore see changes and amendments in the future [24].

1.5 The Role of European Standardisation Organisations and Harmonized European Norms

The 1985 Resolution is to be understood as a high-level endorsement of the principles of the New Approach. It follows the adoption in 1983 of a Directive on technical standards and regulations (Directive 83/189/EEC [25]) which forces Member States and their national standard bodies to exchange information on the adoption of standards and technical regulations ahead of their adoption. While the Directive in itself sets only high-level principles on the exchange of information, the objectives set by the legislator are of importance. Technical regulations adopted at national level are percieved as potential technical barriers to trade and are subject to a review in order to eliminate those that are not strictly necessary. Member States and their standard setting organisations are restricted in their ability to establish new norms in order to allow a free movement of goods between the Member States. This freedom of movement is at the top of the priorities of the legislator. At the same time, the Directive does not prohibit the adoption of norms, nor does it makes standards or technical regulations less relevant. The adoption of such norms at European level is foreseen in the Directive. This is the beginning of a process where standards and norms will become slowly a European matter rather than a purely national one.

The Council resolution of May 1985 defining the New approach insists also on the fact that "*these technical specifications are not mandatory and maintain their status of voluntary standards*". It states therefore that it is up to the manufacturer of consumer products to decide whether or not to use these technical specifications in order to comply with the essential requirements.

Over the time, the process for the establishment of norms, standards and specification at European level has become a solidly established one, taking over from the systems that were (and to a certain extent, still are) in place at national level. A few years later, the European legal framework was complemented by the Council Decision 87/95/EEC [26]. This Decision is specifically focussed on one range of products, called *"information technology and telecommunication"*. It sets the principles of a standardisation policy at European level while taking into consideration the work done already, including at international level (the Decision mentions international bodies such as the European Conference of Postal and Telecommunications Administrations (CEPT). The Decision established in its Article 4 a key principle of the NLF: the Commission is entrusting standard organisations (CEN (European

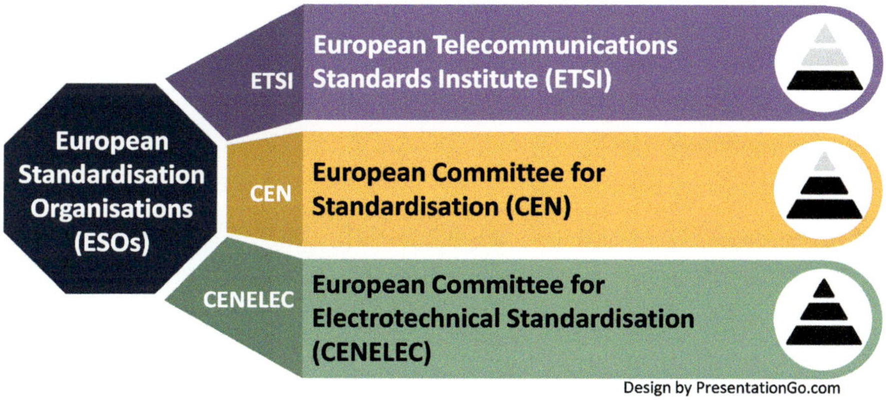

Fig. 1.8 European Standard Organisation (ESOs) as defined by Regulation 2012/1025 [27]

Committee for Standardization), CENELEC (European Committee for Electrotechnical Standardization) and the CEPT are explicitly mentioned) with the task to draw up European standards and norms. The instrument of public procurement at national level is also used in order to promote European standards and norms.

Over the years, the framework of the European standards has been adapted several time, with the latest legal update done through Regulation 2012/1025 [27]. This legislation defines the rules for the adoption of standards at European level. It lists (in its Annex I) three European Standardisation Organisation, or ESOs as illustrated in Fig. 1.8:

The ESOs are required to work in a transparent manner, for instance by publishing an annual work programme (Article 3), to make draft standards available if requested (Article 4), and to ensure the participation of the private sector and civil society (Article 5). The EU institutions are required to also publish an annual programme setting up the priorities (Article 8). But more important, the Regulation sets the rules for the request for standards from the Commission to the ESOs (Article 10), by setting deadlines and by requesting that

> "*European standards and European standardisation deliverables shall be market-driven, take into account the public interest as well as the policy objectives clearly stated in the Commission's request and based on consensus.*"

While the legislation sets high-level principles, it also established the fact that the ESOs are tasked with writing the standards that implement the essential requirements of harmonized legislations. The Regulation reminds that the deliverable of the ESOs allow a presumption of conformity with the harmonized legislations (Recital 5).

The distribution of roles between the actors and the different texts could be summarised at a very high-level as illustrated in Fig. 1.9:

Once the harmonized standards are referenced, they are published in the Official Journal of the European Union and listed in a consolidated database [28]. However, it should be noted that while the lists and detailed reference of standards are published

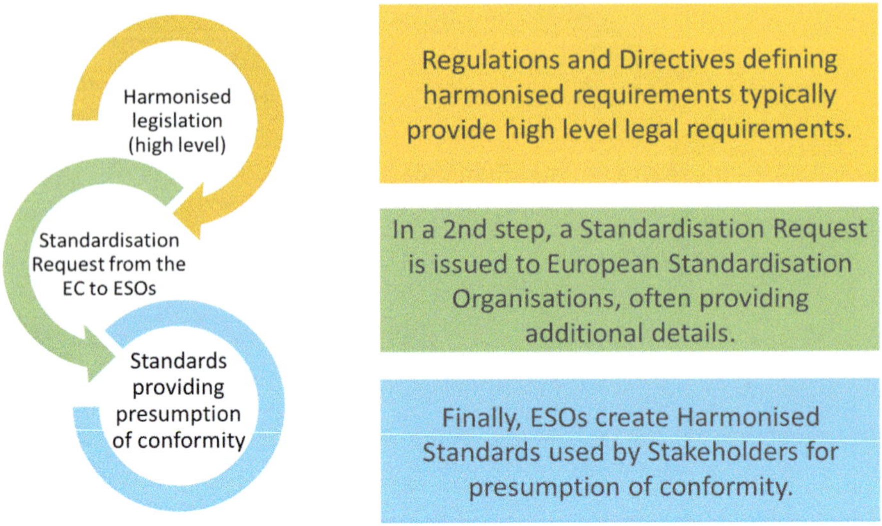

Fig. 1.9 Simplified process from initial harmonized regulation to supporting harmonized standards providing presumption of conformity

in the Official Journal, this does not mean that any economic operator can access the standards immediately. Since standards are developed by the ESOs and given the private nature of ESOs, there is generally a fee to be paid in order to access standards in case of CEN and CENELEC. Only ETSI standards can be accessed free of charge. This has led to discussions over the years, including legal proceedings. If harmonized standards are listed in the Official Journal of the European Union as providing a presumption of conformity with harmonized legislation, should they then be considered part of EU law? And if yes, should they be freely accessible by any citizen? Or should we rather consider that standards, as the result of work of private organisations, is subject to copyright and can therefore only be shared against a fee?

Two court cases in particular have been dealing with the question over the years. In 2016, the European Court of Justice, in the so-called James Elliot case [29], considered itself competent to rule on a standard that was published in the Official Journal of the European Union. It also rules that a harmonized standard to be "*the references to which have been published in the Official Journal of the European Union, forms part of EU law*".

In 2024, another ruling of the European Court of Justice (case Public.Resource.Org and Right to Know [30]) went further and sided with the two parties which had requested from the European Commission an access to harmonized standards. The Commission had argued that harmonized standards were part of EU law but that it should not be forced to release them because there was no overriding interest that would justify it. The Court ruled that in line with the right of access to public document, the disclosure of the harmonized standard requested was justified.

The European Commission is still assessing the consequences of this second ruling, especially regarding the impact on the ESOs who would potentially lose revenues.

1.6 The Blue Guide

In 1994, the European Commission released a document named *Guide to the implementation of directives based on the new approach and the global approach* [31]. This guide is to be understood as "*a manual to be used by practitioners of the New Approach*", according to the European Commission. The colour of the first page of the document gave the name "*Blue Guide*". Over the years, several updates of the guide have been made to adapt it to the new rules in place: in 2000, 2014, 2016 and 2022.

Although it is not a legally binding document, the Blue Guide provides a very helpful explanation of the harmonized and non-harmonized legislations covering consumer products in the EU. For instance, the 2022 version details the concepts that appear in several harmonized legislation, such as "*placing on the market*" and "*making available on the market*". It also explains in details the different modules of the conformity assessment procedures.

1.7 Conclusions

The present chapter introduces the European New Legislative Framework (NLF) and outlines the historic developments that have led to the framework as it is implemented today. The NLF has been proven highly successful and appreciated by stakeholders, since it allows for direct participation of stakeholders to the development of Harmonized European Norms—or Harmonized Standards—determining the implementation of related regulations and thus market access conditions. The corresponding process thus ensures that market access is meeting the requirements of the legislator and also considers implementation constraints by manufacturers and other stakeholders.

This chapter has provided a historical overview of the NLF's development, highlighting its core principles and its impact on the European internal market for goods. Looking ahead, future challenges include ensuring the NLF's continued effectiveness in a globalized marketplace and adapting it to address emerging technologies. The authors trust that the European Commission is actively working to review and update the NLF to maintain its relevance and effectiveness.

Disclaimer Note that the authors have assembled the information above to the best of their knowledge as available at the time when the present book was written. Any interpretation is the personal opinion of the authors. Also, the regulation framework can change and evolve any time. The only legally binding text is the one provided in the documents formally published in the European Union Official Journal.

References

1. 30th anniversary of the single market 1993–2023, European Commission, available at https://single-market-economy.ec.europa.eu/single-market/30th-anniversary_en
2. [1a] Delors, J. (2012). The Single Market, Cornerstone of the EU. Speech on the occasion of the 20th anniversary of the Single Market, Tribune, Notre Europe–Jacqes Delors Institute, 22. https://europe-solidarity.eu/documents/ES_Singlemarket.pdf
3. Council Resolution of 7 May 1985 on a new approach to technical harmonization and standards (OJ C, C/136, 04.06.1985, p. 1, available at https://eur-lex.europa.eu/legal-content/EN/TXT/?uri=CELEX:31985Y0604(01)
4. [2b] Egan, Michelle P., 'Case Studies in Standards Negotiations11', Constructing a European Market: Standards, Regulation, and Governance (Oxford, 2001; online edn, Oxford Academic, 1 Nov. 2003), https://doi.org/10.1093/0199244057.003.0008, accessed 10 Sept. 2024.
5. Market surveillance and product safety, European Paliament, available at https://www.europarl.europa.eu/RegData/bibliotheque/briefing/2013/130491/LDM_BRI%282013%29130491_REV1_EN.pdf
6. The New Approach, European Committee for Standardization, available at https://boss.cen.eu/reference-material/guidancedoc/pages/newapproach/
7. Council Directive 89/336/EEC of 3 May 1989 on the approximation of the laws of the Member States relating to electromagnetic compatibility, available at https://eur-lex.europa.eu/legal-content/EN/ALL/?uri=CELEX%3A31989L0336
8. Council Directive 86/361/EEC of 24 July 1986 on the initial stage of the mutual recognition of type approval for telecommunications terminal equipment, available at https://eur-lex.europa.eu/legal-content/EN/ALL/?uri=CELEX:31986L0361
9. Council Directive 90/385/EEC of 20 June 1990 on the approximation of the laws of the Member States relating to active implantable medical devices, available at https://eur-lex.europa.eu/legal-content/EN/TXT/?uri=CELEX%3A31990L0385
10. Directive 2014/53/EU of the European Parliament and of the Council of 16 April 2014 on the harmonisation of the laws of the Member States relating to the making available on the market of radio equipment and repealing Directive 1999/5/EC Text with EEA relevance, available at https://eur-lex.europa.eu/legal-content/EN/TXT/?uri=celex%3A32014L0053
11. Council Directive 91/263/EEC of 29 April 1991 on the approximation of the laws of the Member States concerning telecommunications terminal equipment, including the mutual recognition of their conformity, available at https://eur-lex.europa.eu/legal-content/EN/TXT/?uri=CELEX%3A31991L0263
12. Directive 98/13/EC of the European Parliament and of the Council of 12 February 1998 relating to telecommunications terminal equipment and satellite earth station equipment, including the mutual recognition of their conformity, available at https://eur-lex.europa.eu/legal-content/EN/TXT/?uri=celex%3A31998L0013
13. Council Directive 93/97/EEC of 29 October 1993 supplementing Directive 91/263/EEC in respect of satellite earth station equipment, available at https://eur-lex.europa.eu/legal-content/EN/TXT/?uri=CELEX%3A31993L0097
14. Directive 1999/5/EC of the European Parliament and of the Council of 9 March 1999 on radio equipment and telecommunications terminal equipment and the mutual recognition of their conformity, available at https://eur-lex.europa.eu/legal-content/EN/TXT/?uri=celex%3A31999L0005
15. New legislative framework, European Commission, available at https://single-market-economy.ec.europa.eu/single-market/goods/new-legislative-framework_en
16. 90/683/EEC: Council Decision of 13 December 1990 concerning the modules for the various phases of the conformity assesment procedures which are intended to be used in the technical harmonization directives, available at https://eur-lex.europa.eu/legal-content/EN/TXT/?uri=CELEX%3A31990D0683

17. 93/465/EEC: Council Decision of 22 July 1993 concerning the modules for the various phases of the conformity assessment procedures and the rules for the affixing and use of the CE conformity marking, which are intended to be used in the technical harmonization directives, available at https://eur-lex.europa.eu/eli/dec/1993/465/oj
18. Decision No 768/2008/EC of the European Parliament and of the Council of 9 July 2008 on a common framework for the marketing of products, and repealing Council Decision 93/465/EEC (Text with EEA relevance), available at https://eur-lex.europa.eu/legal-content/EN/TXT/?uri=CELEX%3A32008D0768
19. Regulation (EC) No 765/2008 of the European Parliament and of the Council of 9 July 2008 setting out the requirements for accreditation and market surveillance relating to the marketing of products and repealing Regulation (EEC) No 339/93 (Text with EEA relevance), available at https://eur-lex.europa.eu/legal-content/en/ALL/?uri=CELEX%3A32008R0765
20. Council Regulation (EEC) No 339/93 of 8 February 1993 on checks for conformity with the rules on product safety in the case of products imported from third countries, available at https://eur-lex.europa.eu/legal-content/EN/ALL/?uri=celex%3A31993R0339
21. Single Market Compliance Space (SMCS), European Commission, available at https://webgate.ec.europa.eu/single-market-compliance-space/home
22. What is accreditation, Dutch Accreditation Council, available at https://www.rva.nl/en/what-is-accreditation
23. Regulation (EU) 2019/1020 of the European Parliament and of the Council of 20 June 2019 on market surveillance and compliance of products and amending Directive 2004/42/EC and Regulations (EC) No 765/2008 and (EU) No 305/2011 (Text with EEA relevance), available at https://eur-lex.europa.eu/eli/reg/2019/1020/oj
24. Industrial products - evaluation of the new legislative framework, European Commission, available at https://ec.europa.eu/info/law/better-regulation/have-your-say/initiatives/12654-Industrial-products-evaluation-of-the-new-legislative-framework_en
25. Council Directive 83/189/EEC of 28 March 1983 laying down a procedure for the provision of information in the field of technical standards and regulations, available at https://eur-lex.europa.eu/legal-content/en/ALL/?uri=CELEX%3A31983L0189
26. 87/95/EEC: Council Decision of 22 December 1986 on standardization in the field of information technology and telecommunications, available at https://eur-lex.europa.eu/legal-content/en/ALL/?uri=CELEX%3A31987D0095
27. Regulation (EU) No 1025/2012 of the European Parliament and of the Council of 25 October 2012 on European standardisation, amending Council Directives 89/686/EEC and 93/15/EEC and Directives 94/9/EC, 94/25/EC, 95/16/EC, 97/23/EC, 98/34/EC, 2004/22/EC, 2007/23/EC, 2009/23/EC and 2009/105/EC of the European Parliament and of the Council and repealing Council Decision 87/95/EEC and Decision No 1673/2006/EC of the European Parliament and of the Council Text with EEA relevance, available at https://eur-lex.europa.eu/eli/reg/2012/1025/oj
28. Harmonised Standards, European Commission, available at https://single-market-economy.ec.europa.eu/single-market/european-standards/harmonised-standards_en
29. James Elliott court case, available at https://curia.europa.eu/juris/document/document.jsf?text=&docid=184891&pageIndex=0&doclang=EN&mode=lst&dir=&occ=first&part=1&cid=9838234
30. Public.Resource.Org and Right to Know court case, available at https://curia.europa.eu/juris/document/document.jsf?text=&docid=283443&pageIndex=0&doclang=EN&mode=lst&dir=&occ=first&part=1&cid=9841601
31. Guide to implementation of Community harmonisation directives based on the new approach and the global approach, European Commission, available at https://op.europa.eu/en/publication-detail/-/publication/3d49c4e8-03de-4a9a-ab41-5d18721eea8a/language-en

Chapter 2
Overview on European Digital Legislations

Markus Mueck, Stephane Du Boispean, and Christophe Gaie

Abstract This chapter provides an overview on the vast space of European Legislations in the Digital Space. A comprehensive summary is provided including an outlook on legislations which are currently in preparation and will likely be upcoming in the future. The set of identified legislations are clustered into a total of 12 categories, including Research and Innovation, Industrial Policy, Connectivity, Data and Privacy and others. For each of those clusters, available legislations are listed and briefly summarized. Also, a link to the full legislation text is provided where available. Obviously, the only legally binding text is the one provided in the documents formally published in the European Union Official Journal. While all such legislationsare important, the subsequent chapters of this book will specifically focus on a selected subset which the authors consider to be of specific importance to stakeholders including users, industry, small and medium-sized enterprises, academia and others. Specific attention will be given on legislations addressing Artificial Intelligence, Cybersecurity, Data Management, Radio Equipment and others. For all stakeholders, it is critical to have a comprehensive understanding of these legislations because they govern the access to the European Union Single Market.

Keywords Regulation · Digital · Artificial intelligence · Cyber security · Radio equipment · Market access · European Union Single Market · Legislation

M. Mueck (✉)
Intel Deutschland GmbH, Neubiberg, Germany
e-mail: Markus.Dominik.Mueck@intel.com

S. Du Boispean
Intel Corporation SA/NV, Kontich, Belgium
e-mail: Stephane.du.Boispean@intel.com

C. Gaie
Head of the Engineering and Digital Innovation Division, Prime Minister Services, Paris, France

M. Mueck and C. Gaie (eds.), *European Digital Regulations*, Intelligent Systems Reference Library 265, https://doi.org/10.1007/978-3-031-80809-8_2

2.1 Introduction

On the level of the European Union, a vast set of European Legislations in the Digital Space have been implemented or are in preparation. The European Commission provides a database where legislation related documents, including European Regulations in the Digital Space, are available [1] and can be downloaded. Previous work outlined the multiple challenges to overcome so as to establish a harmonized digital framework [2]. To reach this objective, the European Think Tank Bruegel [3] has conducted a related analysis and published a corresponding overview [4]. The space of European Legislations in the Digital space is proposed to be clustered in a total of 12 legislation categories which are summarized in Fig. 2.1.

The academic literature also outlines the evolution of the EU legislations towards more regulation, protection of citizens while preserving innovation capabilities [5]. It also pointed out the efforts to make for a better protection of European citizens as well as a necessity to enhance the regulatory framework to reduce equivocal interpretations [6].

In this chapter, we will provide a comprehensive overview on available and upcoming legislations under the above listed categories. While all legislations are important and relevant to stakeholders, subsequent chapters in this book will focus on a key set of legislations that have recently been published or will likely introduce new requirements in the future to be met for ensuring continued access to the European Union Single Market. We indeed expect that especially legislations in the space of Cybersecurity, Data & Privacy, Artificial Intelligence and others will be a game

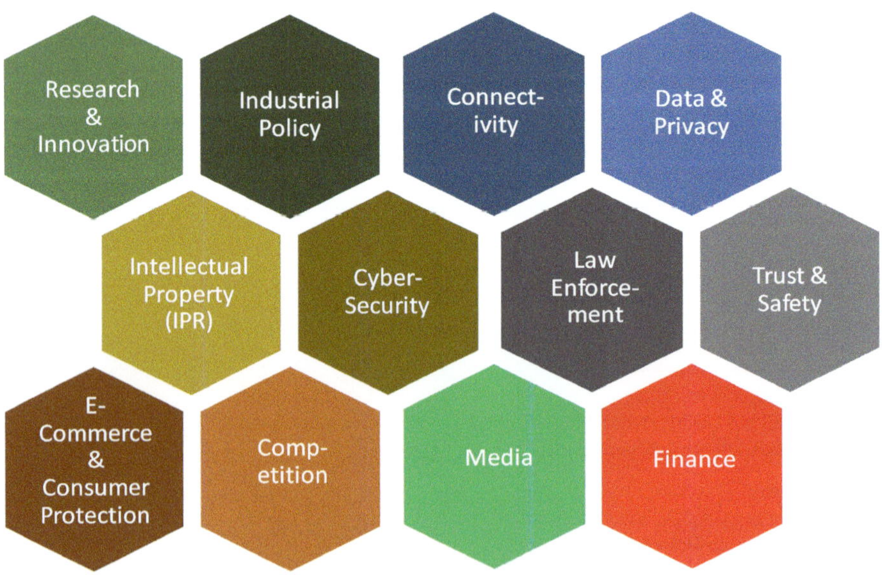

Fig. 2.1 Overview on European Legislations in the Digital space

changer—a substantial number of new market access requirements—so called *essential requirements*—are being introduced which have substantial impact on the design of hardware and software products. Those are of key importance to manufacturers of such products—meeting those requirements and obtaining related compliance certifications are essential, otherwise access to the European Union Single Market is lost.

Note that the information presented throughout the present chapter is evolving—especially the status on planned/ongoing initiatives may change any time. The authors assembled the information below and throughout the book to the best of their knowledge at the time when the present chapter was written. The only legally binding text is the one provided in the documents formally published in the European Union Official Journal.

2.2 Research and Innovation

Existing legislations (i.e., published in the Official Journal of the European Union (OJEU)) governing research and innovation in the European Union are summarized in Table 2.1.

Table 2.1 Existing legislations on research and innovation

Legislations (published in the OJEU)	Summary
Digital Europe Programme Regulation, (EU) 2021/694	The EU launched the Digital Europe Programme for the period 2021–2027, setting out its goals, a seven-year budget, and the funding types and rules
(EU) 2021/694 is available at https://eur-lex.europa.eu/legal-content/EN/TXT/?uri=CELEX%3A32021R0694&qid=1688466008691	
Horizon Europe Regulation, (EU) 2021/695, (EU) 2021/764	Horizon Europe was implemented under the Framework Programme for Research and Innovation with the goal of strengthening the EU's scientific and technological base during the period 2021–2027
(EU) 2021/695 is available at https://eur-lex.europa.eu/legal-content/EN/TXT/?uri=CELEX%3A32021R0695&qid=1688466500998 (EU) 2021/764 is available at https://eur-lex.europa.eu/legal-content/EN/TXT/?uri=CELEX%3A32021D0764	
Regulation on a pilot regime distributed ledger tech. market, (EU) 2022/858	Establishment of a pilot regime for market infrastructures based on distributed ledger technology. This framework provides a basis specifically for crypto-assets including financial instruments
(EU) 2022/858 is available at https://eur-lex.europa.eu/legal-content/EN/TXT/?uri=CELEX%3A32022R0858&qid=1688466540287	

These legislations play a key role to drive innovation and research in the EU by supporting the entire process, from initial ideas to practical applications and testing specific technologies.

As a matter of fact, the Digital Europe Programme Regulation sets up the groundwork by funding research into new technologies and fostering innovation. It focuses on the "idea stage" and development of cutting-edge solutions.

Then, the Horizon Europe Regulation introduces a framework to translate those ideas into practical applications. It bridges the gap between research and real-world implementation.

Finally, the Pilot regime for distributed ledger technology (DLT) establishes regulations and a testing program to assess its viability in a controlled environment. This allows for evaluating its practicality and risks before broader adoption. Other technologies may require specific legislations in the future.

2.3 Industrial Policy

Existing legislations (i.e., published in the Official Journal of the European Union (OJEU)) and planned/ongoing initiatives related to Industrial Policy in the European Union are summarized in Tables 2.2 and 2.3.

These existing legislations represent a significant push by the EU to influence and support various aspects of industry. Two legislations aim to improve financing and investment. First, the Recovery and Resilience Facility & InvestEU provides financial tools for member states and private companies, stimulating investment and economic activity. Second, the European Chips Act and Connecting Europe Facility aim to stimulate investment in semiconductors and infrastructure. Additionally, the High-Performance Computing Joint Undertaking fosters research in key areas, potentially driving industrial applications.

Two legislations aim to define a strategy for industry. First, the Decision on a path to the Digital Decade sets goals for digital advancement, impacting various industries. Second, the legislation concerning Establishing the Strategic Technologies for Europe Platform (STEP) is dedicated to attract investment in technologies deemed crucial for the EU's strategic goals.

These legislations represent a significant effort by the EU to shape its industrial landscape. They focus on financing, infrastructure, research, and strategic sectors but do not tackle the environmental dimension for instance.

While older rules mostly focused on building up industries and investing in them, the new plans aim to help the environment and also support the space industry. The Net Zero Industry Act wants to help industries become greener by investing in technologies that reduce pollution. This fills a gap in the old rules by focusing on how to protect the environment while also growing industries. The EU Space Law is about making space a new important industry, just like the European Chips Act did for chips.

Table 2.2 Existing legislations on industrial policy

Legislations (published in the OJEU)	Summary
Recovery and Resilience Facility Regulation, (EU) 2021/241	The Recovery and Resilience Regulation has set up the framework for a public spending instrument limited in time right after the first wave of the COVID pandemic. It is an instrument of refinancing of public investment projects carried out in the Member States, through loans or grants. In order to finance the instrument, the EU is authorized to borrow money on the markets (by issuing bonds)
(EU) 2021/241 is available at https://eur-lex.europa.eu/legal-content/EN/TXT/?uri=CELEX%3A32021R0241&qid=1688466583043	
InvestEU Programme Regulation, (EU) 2021/523	Establishes the InvestEU Programme, which aims to mobilize private investment in Europe by providing guarantees for loans and equity investments
(EU) 2021/523 is available at https://eur-lex.europa.eu/legal-content/EN/TXT/?uri=CELEX%3A32021R0523&qid=1688466612807	
Connecting Europe Facility Regulation, (EU) 2021/1153	Connecting Europe Facility should be understood as a public investment instrument which defines the framework for EU-level funding of transport, energy and digital infrastructure. The Regulation lays down not only a certain budget but also objectives to be achieved through investments that can be funded under the instrument
(EU) 2021/1153 is available at https://eur-lex.europa.eu/legal-content/EN/TXT/?uri=CELEX%3A32021R1153&qid=1688476625274	
Regulation on High Performance Computing Joint Undertaking, (EU) 2021/1173	The HPC JU is an entity created by this legal instrument in order to support (including through public funding) research and development of High Performance Computing capacities in the EU Also, there is the initiative to open up European supercomputer capacity to AI start-ups. Information is available at https://commission.europa.eu/system/files/2023-10/COM_2023_638_1_EN.pdf
(EU) 2021/1173 is available at https://eur-lex.europa.eu/legal-content/EN/TXT/?uri=CELEX%3A32021R1173&qid=1688476652215	
Regulation on Joint Undertakings under Horizon Europe, (EU) 2021/2085, 2022/0033(NLE)	This Regulation sets the rules for the creation of specific entities (Joint Undertakings) in the context of the Horizon Europe research program
(EU) 2021/2085 is available at https://eur-lex.europa.eu/legal-content/EN/TXT/?uri=CELEX%3A32021R2085&qid=1688478333911 Information on 2022/0033(NLE) is available at https://oeil.secure.europarl.europa.eu/oeil/popups/ficheprocedure.do?reference=2022/0033(COD)&l=en	

(continued)

Table 2.2 (continued)

Legislations (published in the OJEU)	Summary
Decision on a path to the Digital Decade, (EU) 2022/2481	Sets out a roadmap for achieving the EU's Digital Decade goals, which include improving digital skills, connectivity, and public services
(EU) 2022/2481 is available at https://eur-lex.europa.eu/legal-content/EN/TXT/?uri=CELEX%3A32022D2481&qid=1688476685559	
European Chips Act (Regulation), (EU) 2023/1781	This Regulation aims at attracting investments in the design, research and manufacturing of semiconductors in the EU. It addresses funding capacities through different programs and sets a legal framework for the public funding of manufacturing sites. It also sets up intervention mechanisms in case of supply crisis
(EU) 2023/1781 is available at https://eur-lex.europa.eu/legal-content/EN/TXT/?uri=CELEX%3A32023R1781	
European critical raw materials act (Regulation), (EU) 2024/1252	The Critical Raw Material Act aims at making investments in the extraction, processing and recycling of raw materials easier, through expedited permitting and by easing public support to such industrial projects
(EU) 2024/1252 is available at https://eur-lex.europa.eu/eli/reg/2024/1252/oj	
Establishing the Strategic Technologies for Europe Platform (STEP), (EU) 2024/795	The STEP legislation aims at attracting more investments in strategic technologies, listed in the Regulation. It may be seen as overlapping partly with other pieces of legislation in terms of scope but focuses on products that are linked to the political objective of strategic autonomy of the EU
(EU) 2024/795 is available at https://eur-lex.europa.eu/eli/reg/2024/795/oj	

All these new plans show that the EU is working towards a strong industrial policy that helps everyone. They focus on both general growth and specific goals, while also taking care of the environment. But to make this policy even stronger, we should also think about trade deals, competition rules, and how to help people.

2.4 Connectivity

Existing legislations (i.e., published in the Official Journal of the European Union (OJEU)) and planned/ongoing initiatives related to Connectivity in the European Union are summarized in Tables 2.4 and 2.5.

These legislations are dedicated to defining radio spectrum access and usage, reducing broadband deployment costs, and ensuring a fair and secure internet experience for users. Specifically, the Frequency Bands Directive is central for managing

Table 2.3 Planned/ongoing initiatives on industrial policy

Planned/ongoing initiatives	Summary
Net Zero Industry Act, 2023/0081(COD)	The Net Zero Industry Act aims at facilitating investments in the production of so-called Net Zero Technologies, defined by the legislation as technologies enabling the green industrial transition. Its main instruments are measures around public funding and permits for the creation of industrial facilities.
Information on 2023/0081(COD) is available at https://oeil.secure.europarl.europa.eu/oeil/popups/ficheprocedure.do?reference=2023/0081(COD)&l=en	
EU Space Law	The EUSL will focus on ensuring satellite traffic safety, protecting space infrastructures, and guaranteeing long-term space sustainability, collectively forming the EUSL safety, resilience, and sustainability pillars
Information available at https://defence-industry-space.ec.europa.eu/newsroom/consultations/targeted-consultation-eu-space-law_en#:~:text=On%20the%2013th%20of,the%20first%20trimester%20of%202024	

the essential resource for wireless communication technologies like mobile phones, Wi-Fi, and Bluetooth.

The legislations also aim to guarantee fairness of usage and user rights. As a matter of fact, the Open Internet Access Regulation guarantees a level playing field by preventing internet service providers from manipulating internet traffic. Additionally, the European Electronic Communications Code aims to establish a comprehensive set of rules for regulating electronic communications networks, services, and associated resources. It also seeks to stimulate competition, increase investment in new mobile communication technologies, and improve consumer protection.

There also are some additional legislations that offers useful support such as the.eu top-level domain Regulation that uniformizes European online identity by regulating ".eu" domain names, or Roaming Regulation that enables to use a mobile phone seamlessly all aover the EU. These initiatives compose the EU's connectivity strategy. However, gaps still exist to address the process to lay fiber optic cables, or define satellite mobile communications.

While existing legislations focused on a defined set of rules and infrastructure, new initiatives take a more strategic and forward-looking approach, considering future technological changes and EU priorities. Specifically, the ongoing initiative entitled Planning for the Future: RSPP 2.0 will ensure that radio spectrum is available for future technologies. Moreover, the Digital Networks Act will encourage innovation to meet evolving connectivity demands.

Table 2.4 Existing legislations on connectivity

Legislation (published in the OJEU)	Summary
Frequency Bands Directive, (EEC) 1987/372	Allocates radio spectrum bands for various uses (e.g., mobile phone services, Wi-Fi, broadcasting). Consider checking for updates as this directive may be superseded by newer regulations
(EEC) 1987/372 is available at https://eur-lex.europa.eu/legal-content/EN/TXT/?uri=CELEX%3A31987L0372&qid=1688477479232	
Radio Spectrum Decision, (EC) 2002/676	Sets out rules for the use of radio spectrum in the EU (e.g., licensing procedures, power limitations). Similar to Frequency Bands Directive, consider checking for a revised version
(EC) 2002/676 is available at https://eur-lex.europa.eu/eli/dec/2002/676(1)/oj	
Broadband Cost Reduction Directive, (EU) 2014/61, 2023/0046(COD)	Aims to reduce the cost of deploying broadband networks. The original directive (2014) outlines measures like streamlining permitting procedures. It is recommended to consider the proposed amendment (2023/0046(COD))
(EU) 2014/61 is available at https://eur-lex.europa.eu/legal-content/en/TXT/?uri=celex%3A32014L0061 Information on 2023/0046(COD) is available at https://oeil.secure.europarl.europa.eu/oeil/popups/ficheprocedure.do?reference=2023/0046(COD)&l=en	
Open Internet Access Regulation, (EU) 2015/2120	Ensures open and neutral access to the internet for all users, prohibiting internet service providers from throttling or prioritizing specific content
(EU) 2015/2120 is available at https://eur-lex.europa.eu/legal-content/EN/TXT/?uri=CELEX%3A32015R2120	
European Electronic Communications Code Directive (EECC), (EU) 2018/1972	Establishes a framework for regulating electronic communications services in the EU (e.g., phone, internet access), covering areas like consumer rights, network access, and competition

(continued)

Table 2.4 (continued)

Legislation (published in the OJEU)	Summary
(EU) 2018/1972 is available at https://eur-lex.europa.eu/legal-content/EN/TXT/?uri=CELEX%3A32018L1972&qid=1688476963356	
.eu top-level domain Regulation, (EU) 2019/517	Regulates the use of the ".eu" top-level domain for websites, setting eligibility criteria and application procedures for ".eu" domain names
(EU) 2019/517 is available at https://eur-lex.europa.eu/legal-content/EN/TXT/?uri=CELEX%3A32019R0517&qid=1688477111573	
Roaming Regulation, (EU) 2022/612	Allows users to use their mobile phones at no extra cost when traveling within the EU, applying to calls, texts, and data usage
(EU) 2022/612 is available at https://eur-lex.europa.eu/legal-content/EN/TXT/?uri=CELEX%3A32022R0612&qid=1688476996716	
Regulation on the Union Secure Connectivity Programme, (EU) 2023/588	Establishes a program to promote secure and next-generation connectivity infrastructure within the EU, potentially including funding for secure networks, cybersecurity measures, and new technologies
(EU) 2023/588 is available at https://eur-lex.europa.eu/legal-content/EN/TXT/?uri=CELEX%3A32023R0588&qid=1688477074892	

Table 2.5 Planned/ongoing initiatives on connectivity

Planned/ongoing initiatives	Summary
New radio spectrum policy programme (RSPP 2.0)	This initiative will provide strategic planning for spectrum management until at least 2030 by integrating the latest technological developments and achieving EU policy objectives, e.g. • green & digital transition • strategic autonomy • competitiveness • recovery from the COVID-19 crisis
Information available at https://ec.europa.eu/info/law/better-regulation/have-your-say/initiatives/13649-New-radio-spectrum-policy-programme-RSPP-20-_en	
Digital Networks Act	The digital connectivity package aims to start a discussion on concrete proposals with stakeholders related to a set of possible actions to foster the innovation, security and resilience of digital infrastructures
Information available at https://ec.europa.eu/commission/presscorner/detail/en/IP_24_941	

2.5 Data and Privacy

Existing legislations (i.e., published in the Official Journal of the European Union (OJEU)) and planned/ongoing initiatives related to Data & Privacy in the European Union are summarized in Tables 2.6 and 2.7.

The legislations tackle complementary dimensions of data and privacy. First, they focus data protection with the General Data Protection Regulation (GDPR) that is a cornerstone of EU data protection, giving individuals control over their personal data and requiring organizations to ensure these rights. Moreover, the protection of personal is also handled by EU institutions themselves through the Regulation to protect personal data processed by EU institutions.

Then, the legislations define the process to achieve exchange of data through the Regulation on the free flow of non-personal data that facilitates the free movement of non-personal data within the EU, the Open Data Directive that encourages public bodies to make certain data publicly available for reuse, fostering innovation and transparency. The Regulation on data collection and sharing relates to short-term accommodation rental services that details the application on this specific sector.

Other legislations of the domain address the governance of data such as the Data Governance Act that defines a framework for trustworthy data sharing across sectors and borders within the EU, the European Data Act that promotes fairer business practices in the data economy by setting rules for data access and cloud service providers and the Interoperable Europe Act that aims to facilitate the interoperability of digital systems within EU.

Thus, while the EU established many legislations addressing data protection, data flow, governance, and access, there are still rules to define concerning dimensions such as international data transfers or emerging technologies.

Two initiatives are currently engaged to complement the existing legislations. First the Revision of the ePrivacy Regulation aims to strengthen existing protections in electronic communications. Thus, it defines the applicable rules in areas like tracking and cookies, potentially aligning with the GDPR's principles.

Then, the future European Health Data Space Regulation will be dedicated to the construction of a framework for secure access and sharing of health data within the EU, empowering individuals to control their health information. These legislations will reinforce the protection of personal data in specific contexts like online communications and healthcare while enabling data-driven innovation within the EU.

2.6 Intellectual Property Rights (IPR)

Existing legislations (i.e., published in the Official Journal of the European Union (OJEU)) and planned/ongoing initiatives related to Intellectual Property Rights (IPR) in the European Union are summarized in Tables 2.8 and 2.9.

Table 2.6 Existing legislations on data and privacy

Legislation (published in the OJEU)	Summary
European statistics (EC) 2009/223, 2023/0237(COD)	Ensures the coherence and comparability of European statistics
(EC) 2009/223 is available at https://eur-lex.europa.eu/LexUriServ/LexUriServ.do?uri=OJ:L:2009:087:0164:0173:en:PDF Information on 2023/0237(COD) is available at https://oeil.secure.europarl.europa.eu/oeil/popups/ficheprocedure.do?reference=2023/0237(COD)&l=en	
General Data Protection Regulation (GDPR), (EU) 2016/679	Protects the fundamental right to privacy and personal data of individuals in the EU
(EU) 2016/679 is available at https://eur-lex.europa.eu/legal-content/EN/TXT/?uri=CELEX%3A32016R0679&qid=1688477295692	
Regulation to protect personal data processed by EU institutions, bodies, offices and agencies, (EU) 2018/1725	Protects personal data processed by the institutions, bodies, offices, and agencies of the EU
(EU) 2018/1725 is available at https://eur-lex.europa.eu/legal-content/EN/TXT/?uri=CELEX%3A32018R1725&qid=1688477347890	
Regulation on the free flow of non-personal data, (EU) 2018/1807	Enables the free movement of non-personal data within the EU and removes restrictions on such data flows
(EU) 2018/1807 is available at https://eur-lex.europa.eu/legal-content/EN/TXT/?uri=CELEX%3A32018R1807&qid=1689157879252	
Open Data Directive (PSI), (EU) 2019/1024	Promotes the accessibility and re-use of certain public sector information
(EU) 2019/1024 is available at https://eur-lex.europa.eu/legal-content/EN/TXT/?uri=CELEX%3A32019L1024&qid=1688477402478	
Data Governance Act (DGA Regulation), (EU) 2022/868	Establishes a framework for governing data intermediaries and promoting data sharing for innovation
(EU) 2022/868 is available at https://eur-lex.europa.eu/legal-content/EN/TXT/?uri=CELEX%3A32022R0868&qid=1688477427076	
European Data Act (Regulation), (EU) 2023/2854	Establishes rules for fair access to and use of data in certain sectors (likely still under development)
(EU) 2023/2854 is available at https://eur-lex.europa.eu/eli/reg/2023/2854	
Interoperable Europe Act, (EU) 2024/903	Aims to improve interoperability of public services across EU member states. This could impact data exchange by requiring public services to be able to communicate and share data seamlessly
(EU) 2024/903 is available at https://eur-lex.europa.eu/eli/reg/2024/903/oj	

(continued)

Table 2.6 (continued)

Legislation (published in the OJEU)	Summary
Regulation on data collection and sharing relating to short-term accommodation rental services, (EU) 2024/1028	Requires short-term rental platforms (e.g., Airbnb) to share data with authorities for tax and regulatory purposes. This aims to improve tax collection and address potential issues like illegal short-term rentals

(EU) 2024/1028 is available at https://eur-lex.europa.eu/eli/reg/2024/1028/oj

The legislations concerning intellectual property rights (IPR) in the EU cover multiple objectives: to protect the investment in creating databases by granting rights to control access and exploitation, to guarantee the unique appearance of a product, fostering design innovation, and to safeguards confidential business information, crucial for a competitive market.

These legislations have connections with the Copyright Directive, that protects creators of original works like books, music, and software, as well as the Regulation on the unitary patent that creates a single patent system for a large part of the EU, simplifying patent protection.

These initiatives aim to complement the existing legislations by modernizing and clarifying the existing design rights, potentially providing more certainty for businesses, establishing a framework for issuing compulsory licenses in specific situations, potentially balancing patent holder rights with public health or economic needs, and addressing challenges related to patents essential for implementing standardised technologies.

These initiatives demonstrate the EU's ongoing focus on refining the protection of Intellectual Property Rights (IPR). They aim to balance the protection of intellectual property with considerations like access to essential goods, fair licensing practices, and a competitive internal market.

2.7 Cybersecurity

Existing legislations (i.e., published in the Official Journal of the European Union (OJEU)) and planned/ongoing initiatives related to Cybersecurity in the European Union are summarized in Tables 2.10 and 2.11.

These legislations establish a robust cybersecurity framework for the EU. They promote secure ICT products and services, foster collaboration, strengthen critical infrastructure protection, and ensure secure EU institutions [7].

The Cybersecurity Act establishes a system for certifying ICT products, services, and processes that increases the level of cybersecurity within EU. The Regulation to establish a European Cybersecurity Competence Centre creates a central body and a network of national centers (NCCs) to enhance collaboration between member states and improve collective cyber defense capabilities.

Table 2.7 Planned/ongoing initiatives on data and privacy

Planned/ongoing initiatives	Summary
ePrivacy Regulation, 2017/0003(COD)	Protects fundamental rights, including privacy and data protection, in electronic communications. Further ensures the free movement of communications data, equipment, and services within the EU
Information on 2017/0003(COD) is available at https://oeil.secure.europarl.europa.eu/oeil/popups/ficheprocedure.do?reference=2017/0003(COD)&l=en	
European Health Data Space (Regulation), 2022/0140(COD)	Addresses health-specific challenges to electronic health data access and sharing, is one of the priorities of the European Commission in the area of health and will be an integral part of building a European Health Union. EHDS will create a common space where natural persons can easily control their electronic health data
Information on 2022/0140(COD) is available at https://oeil.secure.europarl.europa.eu/oeil/popups/ficheprocedure.do?reference=2022/0140(COD)&l=en	
Harmonisation of GDPR Enforcement, 2023/0202(COD)	Addresses problems in the application of the GDPR in the following areas • Complaints • Procedural rights of parties under investigation • Cooperation and dispute resolution
Information on 2023/0202(COD) is available at https://oeil.secure.europarl.europa.eu/oeil/popups/ficheprocedure.do?reference=2023/0202(COD)&l=en	
Access to vehicle data, functions and resources	Defines conditions for accessing and using such in-vehicle generated data. It aims to enable clear and competition-friendly EU rules for services that are based on access to car data, for example • repair and maintenance • car sharing • mobility as a service • insurance
Information is available at https://ec.europa.eu/info/law/better-regulation/have-your-say/initiatives/13180-Access-to-vehicle-data-functions-and-resources_en	

(continued)

Table 2.7 (continued)

Planned/ongoing initiatives	Summary
GreenData4all	Helps deliver on Europe's green and digital transformation by updating EU rules on environmental geospatial data and on public access to environmental information. The aim is to • enable greater sharing of data between the public & private sectors and with the general public • unlock the full benefits of data sharing for data-driven innovation and evidence-based decisions
Information is available at https://ec.europa.eu/info/law/better-regulation/have-your-say/initiatives/13170-GreenData4All-updated-rules-on-geospatial-environmental-data-and-access-to-environmental-information_en	

Table 2.8 Existing legislations on intellectual property rights (IPR)

Legislation (published in the OJEU)	Summary
Database Directive, (EC) 1996/9	Harmonizes national laws on the legal protection of databases
(EC) 1996/9 is available at https://eur-lex.europa.eu/legal-content/EN/TXT/?uri=celex%3A31996L0009	
Community Design Directive, (EC) 2002/6, 2022/0391(COD)	Protects new designs for products throughout the EU. A design refers to the two-dimensional or three-dimensional appearance of a product
(EC) 2002/6 is available at https://eur-lex.europa.eu/legal-content/EN/ALL/?uri=celex%3A32002R0006 Information on 2022/0391(COD) is available at https://oeil.secure.europarl.europa.eu/oeil/popups/ficheprocedure.do?reference=2022/0391(COD)&l=en	
Enforcement Directive (IPR), (EC) 2004/48	Sets minimum standards for enforcement of intellectual property rights (IPRs) across the EU
(EC) 2004/48 is available at https://eur-lex.europa.eu/legal-content/EN/TXT/?uri=CELEX%3A32004L0048&qid=1688483852608	
Directive on the protection of trade secrets, (EU) 2016/943	Protects confidential business information (trade secrets) from unauthorized disclosure
(EU) 2016/943 is available at https://eur-lex.europa.eu/legal-content/EN/TXT/?uri=CELEX%3A32016L0943&qid=1688483905678	

Table 2.9 Planned/ongoing initiatives on intellectual property rights (IPR)

Planned/ongoing initiatives	Summary
Design Directive, 2022/0392(COD)	The objectives include • modernising and clarifying the existing provisions of the Directive, by amending outdated provisions, increasing legal certainty and detailing design rights in terms of scope and limitations • strengthening interoperability and complementarity of national laws and procedures with the common design system, by adding further substantive rules and introducing principal procedural rules into the Directive in accordance with provisions contained in the Regulation • completing the single market in repair spare parts by introducing a repair clause into the Directive as already contained in the Regulation
Information on 2022/0392(COD) is available at https://oeil.secure.europarl.europa.eu/oeil/popups/ficheprocedure.do?reference=2022/0392(COD)&l=en	
Compulsory licensing of patents, 2023/0129(COD)	The initiative includes the following objectives • enable the EU to rely on compulsory licensing considering the EU crisis instruments • introduce an efficient compulsory licensing scheme with appropriate features to ensure a swift and effective response to crises, maintain a functioning Internal Market, and guarantee the supply and free movement of crisis-critical products subject to compulsory licensing within the EU
Information on 2023/0129(COD) is available at https://oeil.secure.europarl.europa.eu/oeil/popups/ficheprocedure.do?reference=2023/0129(COD)&l=en	
Standard essential patents, 2023/0133(COD)	The overall objectives of this proposed initiative are to (i) ensure that end users, including small businesses and EU consumers benefit from products based on the latest standardised technologies (ii) make the EU attractive for standards innovation and (iii) The overall objectives of this proposed initiative are to: encourage both Standards Essential Patents (SEP) holders and implementers to innovate in the EU, make and sell products in the EU and be competitive in non-EU markets
Information on 2023/0133(COD) is available at https://oeil.secure.europarl.europa.eu/oeil/popups/ficheprocedure.do?reference=2023/0133(COD)&l=en	

The NIS 2 Directive strengthens cybersecurity requirements for “essential and important entities” in critical sectors (energy, transport, etc.) while the Cybersecurity Regulation sets specific cybersecurity measures for EU institutions themselves to reduce vulnerability of EU Member States against cyberattacks.

These initiatives aim to complement the existing legislations towards cybersecurity. The Information Security Regulation describes information security practices and foster a more coherent security culture within the EU administration. The Cyber Resilience Act targets manufacturers of products with digital elements (hardware and software) and aim to ensure these products are more secure throughout their

Table 2.10 Existing legislations on cybersecurity

Legislation (published in the OJEU)	Summary
Regulation for a Cybersecurity Act, (EU) 2019/881 2023/0108(COD)	Establishes a framework for the EU-wide certification of ICT products, services, and processes to ensure a common level of cybersecurity
(EU) 2019/881 is available at https://eur-lex.europa.eu/legal-content/EN/TXT/?uri=CELEX%3A32019R0881&qid=1689158188697 Information on 2023/0108(COD) is available at https://oeil.secure.europarl.europa.eu/oeil/popups/ficheprocedure.do?reference=2023/0108(COD)&l=en	
Regulation to establish a European Cybersecurity Competence Centre, (EU) 2021/887	Establishes the European Cybersecurity Competence Centre and a Network of National Coordination Centres (NCCs) to foster cybersecurity cooperation and expertise across the EU
(EU) 2021/887 is available at https://eur-lex.europa.eu/legal-content/EN/TXT/?uri=CELEX%3A32021R0887&qid=1689158232142	
NIS 2 Directive, (EU) 2022/2555	Strengthens cybersecurity requirements for essential and important entities across various sectors (e.g., energy, transport, waste management and others)
(EU) 2022/2555 is available at https://eur-lex.europa.eu/legal-content/EN/TXT/?uri=CELEX%3A32022L2555&qid=1688480129686	
Cybersecurity Regulation, (EU) 2023/2841	Sets out specific cybersecurity measures for EU institutions, bodies, offices, and agencies
(EU) 2023/2841 is available at https://eur-lex.europa.eu/eli/reg/2023/2841	

lifecycle. The Cyber Solidarity Act outlines EU-wide cooperation and response to cyber threats.

The EU is making significant efforts to strengthen its cybersecurity posture through a focus on internal EU security, product security, and EU-wide cooperation.

2.8 Law Enforcement

Existing legislations (i.e., published in the Official Journal of the European Union (OJEU)) and planned/ongoing initiatives related to Law Enforcement in the European Union are summarized in Tables 2.12 and 2.13.

These legislations specify various aspects of law enforcement in the EU, including data use, tackling specific crimes, information sharing, and evidence collection. Further initiatives should complement the domain by developing continuous monitoring, fostering cross-border cooperation, and ensuring a balance with fundamental rights.

Table 2.11 Planned/ongoing initiatives on cybersecurity

Planned/ongoing initiatives	Summary
Information Security Regulation, 2022/0084(COD)	This initiative seeks to simplify the different legal frameworks of the Union institutions and bodies by • Establishing standardized information (i.e. with harmonised and comprehensive categories), as well as common handling rules for all Union institutions and bodies • Setting up a lean cooperation scheme on information security between Union institutions and bodies. It offers them the opportunity to develop aunified approach to information security across the European administration • Modernising information security policies across all levels for Union institutions and bodies, considering digital transformation and teleworking
Information on 2022/0084(COD) is available at https://eur-lex.europa.eu/legal-content/EN/TXT/?uri=CELEX%3A52022PC0119	
Cyber Resilience Act, 2022/0272(COD)	Objectives to ensure the proper functioning of the internal market (1) foster the development of secure products with digital elements by ensuring that hardware and software products are launched with fewer vulnerabilities and ensure that manufacturers prioritize security throughout their product lifecycle and (2) enable users to consider cybersecurity when choosing and using products with digital components. Four specific objectives were identified (i) ensure that manufacturers improve the security of products with digital elements from the design and development phase and throughout the lifecycle (ii) ensure a coherent cybersecurity framework, facilitating compliance for hardware and software producers (iii) enhance the transparency of security properties of products with digital elements and (iv) enable businesses and consumers to use products with digital elements securely
Information on 2022/0272(COD) is available at https://oeil.secure.europarl.europa.eu/oeil/popups/ficheprocedure.do?reference=2022/0272(COD)&l=en	

(continued)

Table 2.11 (continued)

Planned/ongoing initiatives	Summary
Cyber Solidarity Act (Regulation), 2023/0109(COD)	Reinforce solidarity at Union level in order to enhance the ability to detect, prepare for and respond to cybersecurity threats and incidents through the following specific objectives • to improve coordinated EU efforts to detect and monitor cyber threats and incidents, thereby fostering European technological leadership in cybersecurity • to enhance the preparedness of critical entities across the EU and foster solidarity by developing shared capabilities to respond to major cybersecurity incidents, including offering incident response support to third countries associated with the Digital Europe Programme (DEP) • to strengthen the Union's ability to respond effectively to significant or large-scale incidents by reviewing and analyzing them, identifying key lessons, and making recommendations where necessary

Information on 2023/0109(COD) is available at https://oeil.secure.europarl.europa.eu/oeil/popups/ficheprocedure.do?reference=2023/0109(COD)&l=en

These planned initiatives, the Directive on combating violence against women and the digitalization of travel documents, would complement existing legislations by addressing a specific crime type (violence against women) and potentially improving border security through travel document digitalization.

2.9 Trust and Safety

Existing legislations (i.e., published in the Official Journal of the European Union (OJEU)) and planned/ongoing initiatives related to Trust & Safety in the European Union are summarized in Tables 2.14 and 2.15.

The European AI Act is a major European initiative to regulate the usage of AI technologies, with a specific focus on certain General Purpose AI (GPAI) systems and High Risk AI systems. Further legislations cover the domain of Product Safety and Consumer Protection with the Product Liability Directive, the Toys Regulation, the General Product Safety Regulation, the Machinery Regulation as well as the European Standardization Regulation. It also details the topic of Digital Trust and Security through the revision of the eIDAS Regulation and the Radio Equipment Directive.

While these legislations create a significant framework for trust and safety in the EU, there are still dimensions to cover in online safety and ensuring consistent enforcement across member states.

The mentioned initiatives will complement the current legislations as follows. The AI Liability Directive address potential responsibilities associated with emerging

Table 2.12 Existing legislations on law enforcement

Legislation (published in the OJEU)	Summary
Law Enforcement Directive, (EU) 2016/680	Sets out rules for processing personal data by law enforcement authorities for crime prevention, investigation, prosecution, and punishment
(EU) 2016/680 is available at https://eur-lex.europa.eu/legal-content/EN/TXT/?uri=celex%3A32016L0680	
Directive on combating fraud and counterfeiting of non-cash means of payment, (EU) 2019/713	Criminalizes certain activities related to fraud and counterfeiting of non-cash payment methods (e.g., credit cards)
(EU) 2019/713 is available at https://eur-lex.europa.eu/legal-content/EN/TXT/?uri=CELEX%3A32019L0713	
Regulation on interoperability between EU information systems in the field of borders and visa, (EU) 2019/817	Improves the Union's data management architecture for border management and security
(EU) 2019/817 is available at https://eur-lex.europa.eu/legal-content/EN/TXT/?uri=celex%3A32019R0817	
Regulation on terrorist content online, (EU) 2021/784	Imposes obligations on hosting providers and online platforms to take down terrorist content within a specific timeframe
(EU) 2021/784 is available at https://eur-lex.europa.eu/legal-content/EN/TXT/?uri=CELEX%3A32021R0784&qid=1688480356330	
Temporary CSAM Regulation, (EU) 2021/1232, 2022/0155(COD)	Establishes temporary measures allowing for the detection, reporting, removal, and blocking of online child sexual abuse material (CSAM)
(EU) 2021/1232 is available at https://eur-lex.europa.eu/legal-content/EN/TXT/?uri=celex%3A32021R1232 Information on 2022/0155(COD) is available at https://oeil.secure.europarl.europa.eu/oeil/popups/ficheprocedure.do?reference=2022/0155(COD)&l=en	
E-evidence Regulation, (EU) 2023/1543	Establishes procedures for collecting, handling, and admitting electronic evidence (e.g., emails, digital documents) in criminal investigations and proceedings
(EU) 2023/1543 is available at https://eur-lex.europa.eu/legal-content/EN/TXT/?uri=CELEX%3A32023R1543	

AI technologies. The Eco-design Regulation promotes a more sustainable product lifecycle, potentially contributing to overall product safety.

The combination of those legislations and initiatives broadly covers the topic of trust and safety. Indeed, it details the traditional product safety concerns, emerging AI risks, and environmental considerations, creating a more comprehensive framework for trust in the EU.

Table 2.13 Planned/ongoing initiatives on law enforcement

Planned/ongoing initiatives	Summary
Directive on combating violence against women, 2022/0066(COD)	Aims to effectively combat violence against women and domestic violence throughout the EU. It does so by proposing measures in the following areas: the criminalisation of and sanctions for relevant offences; protection of victims and access to justice; victim support; prevention; coordination and cooperation
Information on 2022/0066(COD) is available at https://oeil.secure.europarl.europa.eu/oeil/popups/ficheprocedure.do?reference=2022/0066(COD)&l=en	
Digitalization of travel documents	As announced in the Schengen strategy adopted in 2021, this initiative aims to decide on a common format for digital travel documents and to make travelling easier
Information available at https://ec.europa.eu/info/law/better-regulation/have-your-say/initiatives/13514-Travel-digitalising-travel-documents-to-make-travelling-easier_en	

2.10 E-Commerce and Consumer Protection

Existing legislations (i.e., published in the Official Journal of the European Union (OJEU)) and planned/ongoing initiatives related to E-Commerce & Consumer Protection in the European Union are summarized in Tables 2.16 and 2.17.

This combination of legislations provides a robust framework for E-Commerce and Consumer Protection in the EU. It protects consumers from unfair practices, ensures transparency, and promotes a safe and fair digital environment. However, continuous review and adaptation might be necessary to address emerging technologies.

The new initiatives aim to improve consumer protection and enforce a right to repair that addresses. Both initiatives contribute to a more comprehensive framework for E-commerce and Consumer Protection in the EU.

2.11 Competition

Existing legislations (i.e., published in the Official Journal of the European Union (OJEU)) and planned/ongoing initiatives related to Competition in the European Union are summarized in Tables 2.18 and 2.19.

Table 2.14 Existing legislations on trust and safety

Legislation (published in the OJEU)	Summary
AI Act (Regulation), (EU) 2024/1689	Proposes a set of rules to govern the development, deployment, and use of Artificial Intelligence (AI) systems. Specific requirements are introduced for certain General Purpose AI (GPAI) systems and High Risk AI systems
(EU) 2024/1689 is available at https://eur-lex.europa.eu/eli/reg/2024/1689/oj	
Product Liability Directive (PLD), (EEC) 1985/374, 2022/0302(COD)	The PLD revision is supposed to modernise the existing EU level framework which dates back from 1985. It introduces several changes in the conditions under which consumers can introduce legal actions against product manufacturers. The rationale behind the revision is that the legal framework should, according to the Commission, be updated in order to take the digital revolution into account
(EEC) 1985/374 is available at https://eur-lex.europa.eu/legal-content/EN/TXT/?uri=celex%3A31985L0374 Information on 2022/0302(COD) is available at https://oeil.secure.europarl.europa.eu/oeil/popups/ficheprocedure.do?lang=en&reference=2022/0302(COD)	
Toys Regulation, (EC) 2009/48, 2023/0290(COD)	The new Toys Regulation introduces changes in the requirements linked to the use of chemical products in toys
(EC) 2009/48 is available at https://eur-lex.europa.eu/legal-content/EN/TXT/?uri=CELEX%3A02009L0048-20221205 Information on 2023/0290(COD) is available at https://eur-lex.europa.eu/legal-content/EN/TXT/?uri=CELEX%3A02009L0048-20221205	
European Standardization Regulation, (EU) 2012/1025	Sets out the framework for European standardization organizations (e.g., CEN, CENELEC) to develop harmonized standards for products
(EU) 2012/1025 is available at https://eur-lex.europa.eu/legal-content/EN/TXT/?uri=CELEX%3A02012R1025-20230709	
eIDAS Regulation, (EU) 2014/910, 2021/0136(COD)	Establishes a framework for electronic identification and trust services (e.g., digital signatures, electronic seals)

(continued)

Table 2.14 (continued)

Legislation (published in the OJEU)	Summary
(EU) 2014/910 is available at https://eur-lex.europa.eu/legal-content/EN/TXT/?uri=CELEX%3A32014R0910&qid=1689159082516 Information on 2021/0136(COD) is available at https://oeil.secure.europarl.europa.eu/oeil/popups/ficheprocedure.do?reference=2021/0136(COD)&l=en	
Radio Equipment Directive (RED), (EU) 2014/53	Regulates the placing on the market of radio equipment to ensure electromagnetic compatibility and safety
(EU) 2014/53 is available at https://eur-lex.europa.eu/legal-content/EN/TXT/?uri=CELEX%3A02014L0053-20221227	
Regulation for a Single Digital Gateway, (EU) 2018/1724	Establishes a Single Digital Gateway for citizens and businesses to access online procedures, information, and assistance across the EU
(EU) 2018/1724 is available at https://eur-lex.europa.eu/legal-content/EN/TXT/?uri=celex%3A32018R1724	
General Product Safety Regulation, (EU) 2023/988	Establishes a stronger framework for ensuring the safety of all products placed on the EU market
(EU) 2023/988 is available at https://eur-lex.europa.eu/legal-content/EN/TXT/?uri=CELEX%3A32023R0988&qid=1688481124734	
Machinery Regulation, (EU) 2023/1230	Updates and strengthens the rules for the design, manufacture, and placing on the market of machinery
(EU) 2023/1230 is available at https://eur-lex.europa.eu/legal-content/EN/TXT/?uri=CELEX%3A32023R1230	

Table 2.15 Planned/ongoing initiatives on trust and safety

Planned/ongoing initiatives	Summary
Eco-design Regulation, 2022/0095(COD)	Proposes new requirements for the environmental performance of products throughout their lifecycle (e.g., energy efficiency, resource use, recyclability)
Information on 2022/0095(COD) is available at https://oeil.secure.europarl.europa.eu/oeil/popups/ficheprocedure.do?reference=2022/0095(COD)&l=en	
AI Liability Directive, 2022/0303(COD)	Aims to complement the AI Act by establishing specific rules on who is liable (e.g., manufacturer, user, operator) for damages caused by AI systems
Information on 2022/0303(COD) is available at https://oeil.secure.europarl.europa.eu/oeil/popups/ficheprocedure.do?reference=2022/0303(COD)&l=en	

The legislations define the rules to ensure fair competition within EU. They address mergers, anti-competitive agreements, fair trading practices, unfair competition from gatekeepers and foreign subsidies, and promote a well-functioning market. However, continuous monitoring, enforcement efforts, and adaptation to emerging issues might be necessary to ensure a level playing field and fair competition in the EU.

Table 2.16 Existing legislations on E-commerce and consumer protection

<table>
<tr><th>Legislation (published in the OJEU)</th><th>Summary</th></tr>
<tr><td>Unfair Contract Terms Directive (UCTD), (EEC) 1993/13</td><td>Protects consumers from unfair terms in standard contracts imposed by businesses</td></tr>
<tr><td colspan="2">(EEC) 1993/13 is available at https://eur-lex.europa.eu/legal-content/EN/TXT/?uri=CELEX%3A31993L0013&qid=1688481881146</td></tr>
<tr><td>Price Indication Directive; (EC) 1998/6</td><td>Require sellers to display the selling price and price per unit of products to enhance consumer knowledge and enable easier price comparisons</td></tr>
<tr><td colspan="2">(EC) 1998/6 is available at https://eur-lex.europa.eu/legal-content/EN/TXT/?uri=CELEX%3A01998L0006-20220528</td></tr>
<tr><td>E-commerce Directive, (EC) 2000/31</td><td>Establishes rules for electronic commerce (online shopping) within the EU</td></tr>
<tr><td colspan="2">(EC) 2000/31 is available at https://eur-lex.europa.eu/legal-content/EN/TXT/?uri=CELEX%3A32000L0031&qid=1688481922427</td></tr>
<tr><td>Unfair Commercial Practices Directive (UCPD), (EC) 2005/29</td><td>Protects consumers from unfair commercial practices used by businesses</td></tr>
<tr><td colspan="2">(EC) 2005/29 is available at https://eur-lex.europa.eu/legal-content/EN/TXT/?uri=CELEX%3A32005L0029&qid=1688481967547</td></tr>
<tr><td>Directive on Consumer Rights (CRD), (EU) 2011/83, 2022/0147(COD)</td><td>Strengthens consumer rights for various purchases (online and offline)</td></tr>
<tr><td colspan="2">(EU) 2011/83 is available at https://eur-lex.europa.eu/legal-content/EN/TXT/?uri=CELEX%3A32011L0083&qid=1688482023928
Information on 2022/0147(COD) is available at https://oeil.secure.europarl.europa.eu/oeil/popups/ficheprocedure.do?reference=2022/0147(COD)&l=en</td></tr>
<tr><td>e-invoicing Directive, (EU) 2014/55</td><td>Promotes the use of electronic invoicing in public procurement</td></tr>
<tr><td colspan="2">(EU) 2014/55 is available at https://eur-lex.europa.eu/legal-content/EN/TXT/?uri=CELEX%3A32014L0055&qid=1688482072077</td></tr>
<tr><td>Geo-Blocking Regulation, (EU) 2018/302</td><td>Prohibits geo-blocking (discrimination based on geographic location) for online goods and services within the EU</td></tr>
<tr><td colspan="2">(EU) 2018/302 is available at https://eur-lex.europa.eu/legal-content/EN/TXT/?uri=CELEX%3A32018R0302&qid=1689159339203</td></tr>
<tr><td>Regulation on cooperation for the enforcement of consumer protection laws, (EU) 2017/2394</td><td>Introduce new measures to effectively address the enforcement challenges of the Single Market, including the challenges of the Digital Single Market</td></tr>
<tr><td colspan="2">(EU) 2017/2394 is available at https://eur-lex.europa.eu/eli/reg/2017/2394/oj</td></tr>
<tr><td>Digital content Directive, (EU) 2019/770</td><td>Complements Consumer Rights Directive for digital content like e-books, music, and games</td></tr>
<tr><td colspan="2">(EU) 2019/770 is available at https://eur-lex.europa.eu/legal-content/EN/TXT/?uri=CELEX%3A32019L0770&qid=1688482216999</td></tr>
</table>

(continued)

Table 2.16 (continued)

Legislation (published in the OJEU)	Summary
Directive on certain aspects concerning contracts for the sale of goods, (EU) 2019/771	Updates rules for consumer goods contracts, including conformity assessment and remedies
(EU) 2019/771 is available at https://eur-lex.europa.eu/legal-content/EN/TXT/?uri=CELEX%3A32019L0771&qid=1688482240672	
Digital Services Act (DSA Regulation), (EU) 2022/2065	Regulates the responsibilities of online platforms (e.g., social media, marketplaces) to tackle illegal and harmful content
(EU) 2022/2065 is available at https://eur-lex.europa.eu/legal-content/EN/TXT/?uri=CELEX%3A32022R2065&qid=1688482264794	
Political Advertising Regulation, (EU) 2024/900	This Regulation aims to ensure that the provision of political advertising is in full respect of fundamental rights
(EU) 2024/900 is available at https://eur-lex.europa.eu/legal-content/EN/TXT/?uri=CELEX%3A32024R0900&qid=1714379247825	

Table 2.17 Planned/ongoing initiatives on E-commerce and consumer protection

Planned/ongoing initiatives	Summary
Right to repair Directive, 2023/0083(COD)	Consumers often do not seek to repair defective products, but discard them prematurely, even though they could be repaired and used for longer. This regulation introduces an obligation for producers to repair defects outside the liability of the seller upon the request of consumers and against a price
Information on 2023/0083(COD) is available at https://commission.europa.eu/document/afb20917-5a6c-4d87-9d89-666b2b775aa1_en	
Multimodal digital mobility services (MDMS)	Traveling across Europe using various modes of transport can be complex. Many people rely on digital tools to plan their journeys and purchase tickets. This initiative seeks to improve the connection between public transport and rail services, creating a seamless multimodal travel experience and supporting the EU Green Deal
Information is available at https://ec.europa.eu/info/law/better-regulation/have-your-say/initiatives/13133-Multimodal-digital-mobility-services_en	
Consumer protection: strengthened enforcement cooperation	The digitalisation of consumer markets and new business models pose new challenges for enforcing EU consumer law. Consumer authorities and the Commission need to cooperate closely to effectively address non-compliant business practices across borders. A flexible, deterrent, cost effective and rapid response is key, in particular where breaches are widespread or affect entire sectors. This initiative makes targeted changes to the Consumer Protection Cooperation Regulation to address these issues
Information is available at https://ec.europa.eu/info/law/better-regulation/have-your-say/initiatives/13535-Consumer-protection-strengthened-enforcement-cooperation_en	

Table 2.18 Existing legislations on competition

Legislation (published in the OJEU)	Summary
EC Merger regulation, (EC) 2004/139, update soon	The primary legal documents for merger decisions are the EC Merger Regulation and the Implementing Regulation. The Merger Regulation outlines the key rules for evaluating mergers, while the Implementing Regulation addresses procedural matters like notifications, deadlines, and hearing rights
(EC) 2004/139 is available at https://eur-lex.europa.eu/legal-content/EN/ALL/?uri=celex%3A32004R0139	
Technology Transfer Block Exemption, (EC) 2014/316	Exempts certain types of technology transfer agreements (e.g., licenses, know-how) from the EU's competition rules
(EC) 2014/316 is available at https://eur-lex.europa.eu/legal-content/EN/TXT/?uri=CELEX%3A32014R0316	
Company Law Directive, (EU) 2017/1132, 2023/0089(COD)	Harmonizes rules on company law for various types of businesses (e.g., public limited companies, private companies) across the EU. (Proposal for revision included)
(EU) 2017/1132 is available at https://eur-lex.europa.eu/legal-content/EN/TXT/?uri=celex%3A32017L1132 Information on 2023/0089(COD) is available at https://oeil.secure.europarl.europa.eu/oeil/popups/ficheprocedure.do?reference=2023/0089(COD)&l=en	
Market Surveillance Regulation, (EU) 2019/1020	Establishes a framework for market surveillance of products placed on the EU market to ensure compliance with relevant safety and regulatory requirements
(EU) 2019/1020 is available at https://eur-lex.europa.eu/legal-content/EN/TXT/?uri=CELEX%3A32019R1020&qid=1688482830025	
P2B Regulation, (EU) 2019/1150	Fair Trading Practices Directive (Business to Business)
(EU) 2019/1150 is available at https://eur-lex.europa.eu/legal-content/EN/TXT/?uri=CELEX%3A32019R1150&qid=1688482861768	
Single Market Programme, (EU) 2021/690	The Single Market Programme (SMP) is the EU funding programme to help the single market reach its full potential and ensure Europe's recovery from the COVID-19 pandemic
(EU) 2021/690 is available at https://eur-lex.europa.eu/legal-content/en/TXT/?uri=CELEX%3A32021R0690	
Vertical Block Exemption Regulation (VBER), (EU) 2022/720	Exempts certain types of vertical agreements (between companies at different levels of the supply chain) from the EU's competition rules
(EU) 2022/720 is available at https://eur-lex.europa.eu/legal-content/EN/TXT/?uri=CELEX%3A32022R0720&qid=1688483379041	
Digital Market Act (DMA Regulation), (EU) 2022/1925	Regulates the behavior of large gatekeeper platforms (e.g., search engines, social media) to promote fair competition in the digital market

(continued)

Table 2.18 (continued)

Legislation (published in the OJEU)	Summary
(EU) 2022/1925 is available at https://eur-lex.europa.eu/legal-content/EN/TXT/?uri=CELEX%3A32022R1925&qid=1688483404069	
Regulation on distortive foreign subsidies, (EU) 2022/2560	Empowers the EU to investigate and address foreign subsidies that distort competition within the EU market
(EU) 2022/2560 is available at https://eur-lex.europa.eu/legal-content/EN/TXT/?uri=CELEX%3A32019L0771&qid=1688482240672	
Horizontal Block Exemption Regulations (HBER), (EU) 2023/1066, (EU) 2023/1067	Exempt certain types of horizontal agreements (between companies at the same level of the supply chain) from the EU's competition rules
(EU) 2023/1066 is available at https://eur-lex.europa.eu/eli/reg/2023/1066/oj (EU) 2023/1067 is available at https://eur-lex.europa.eu/eli/reg/2023/1067/oj	

Table 2.19 Planned/ongoing initiatives on competition

Planned/ongoing initiatives	Summary
Platform Work Directive, 2021/0414(COD)	Aims to improve working conditions and social rights for platform workers (e.g., those working for online platforms like ride-hailing or food delivery services)
Information on 2021/0414(COD) is available at https://oeil.secure.europarl.europa.eu/oeil/popups/ficheprocedure.do?reference=2021/0414(COD)&l=en	
Single Market Emergency Instrument (SMEI), 2022/0278(COD)	Provides a mechanism for the EU to take safeguard measures in response to serious threats to the functioning of the Single Market
Information on 2022/0278(COD) is available at https://oeil.secure.europarl.europa.eu/oeil/popups/ficheprocedure.do?reference=2022/0278(COD)&l=en	

These initiatives would complement existing competition law. The Platform Work Directive could indirectly affect competition by improving platform worker rights and potentially leveling the playing field. The SMEI provides a safety net to address unforeseen threats to the Single Market, potentially including competition distortions.

2.12 Media

Existing legislations (i.e., published in the Official Journal of the European Union (OJEU)) and planned/ongoing initiatives related to Media in the European Union are summarized in Tables 2.20 and 2.21.

The legislations of Media establish a framework for regulating media in the EU, covering traditional broadcast media, online platforms offering audio-visual content,

Table 2.20 Existing legislations on media

Legislation (published in the OJEU)	Summary
Satellite and Cable I Directive, (EEC) 1993/83	Establishes minimum rules for the regulation of cable television and satellite broadcasts transmitted from EU member states
(EEC) 1993/83 is available at https://eur-lex.europa.eu/legal-content/EN/TXT/?uri=CELEX%3A31993L0083&qid=1688484019518	
Information Society Directive, (EC) 2001/29	This Directive concerns the legal protection of copyright and related rights in the framework of the internal market, with particular emphasis on the information society
(EC) 2001/29 is available at https://eur-lex.europa.eu/legal-content/EN/TXT/?uri=CELEX%3A02001L0029-20190606&qid=1688484065665	
Audio-visual Media Services Directive (AVMSD), (EU) 2010/13	Regulates the provision of audio-visual media services (e.g., television broadcasting, video-on-demand platforms) in the EU
(EU) 2010/13 is available at https://eur-lex.europa.eu/legal-content/EN/TXT/?uri=CELEX%3A02010L0013-20181218&qid=1688484163571	
Portability Regulation, (EU) 2017/1128	Enables consumers to easily switch electronic communications service providers (e.g., phone, internet) while keeping their phone numbers
(EU) 2017/1128 is available at https://eur-lex.europa.eu/legal-content/EN/TXT/?uri=CELEX%3A02017R1128-20170630&qid=1688484195904	
Satellite and Cable II Directive, (EU) 2019/789	Complements and updates the Satellite and Cable I Directive to address the evolving media landscape, including online transmission
(EU) 2019/789 is available at https://eur-lex.europa.eu/legal-content/EN/TXT/?uri=CELEX%3A32019L0789&qid=1688484223814	
Copyright Directive, (EU) 2019/790	Introduces new rights and obligations for online platforms and rightholders (e.g., content creators) in the digital copyright framework
(EU) 2019/790 is available at https://eur-lex.europa.eu/legal-content/EN/TXT/?uri=CELEX%3A32019L0790&qid=1688484249021	

consumer rights, and copyright. However, continuous monitoring and ensuring effective enforcement across the EU are necessary for a comprehensive approach to media regulation.

The planned initiatives, the European Media Freedom Act and the initiative on remuneration of non-EU musicians, would strengthen the existing media regulatory framework in the EU.

Table 2.21 Planned/ongoing initiatives on media

Planned/ongoing initiatives	Summary
European Media Freedom Act, 2022/0277(COD)	Aims to protect and promote media freedom and pluralism within the EU
Information on 2022/0277(COD) is available at https://oeil.secure.europarl.europa.eu/oeil/popups/ficheprocedure.do?reference=2022/0277(COD)&l=en	
Remuneration of musicians from third countries for recorded music played in the EU	This initiative will amend the EU's rules on rental and lending rights and on certain rights related to copyright. The aim is to clarify certain conditions and ensure legal certainty on the remuneration of music performers and record producers from third countries when their records are played in the EU on radio or TV, or in public places such as bars and restaurants
Information is available at https://ec.europa.eu/info/law/better-regulation/have-your-say/initiatives/13530-Remuneration-of-music-performers-and-record-producers-from-third-non-EU-countries-for-recorded-music-played-in-the-EU_en	

2.13 Finance

Existing legislations (i.e., published in the Official Journal of the European Union (OJEU)) and planned/ongoing initiatives related to Finance in the European Union are summarized in Tables 2.22 and 2.23.

The legislations concerning Finance cover taxation, administrative cooperation, payments, financial sector resilience, and crypto-assets. However, they need to be complemented for continuous monitoring, adaptation to address new financial technologies, and ensuring effective enforcement are crucial for a comprehensive approach to financial regulation in the EU [8].

The planned initiatives are expected to expand the scope of EU financial legislation as follows: Financial Data Access fosters innovation by enabling new financial services, Payment Services will modernize the framework and adapts to evolving trends, the Digital Euro is expected to introduce a new payment instrument with potential implications for the financial system, and the Regulation on Combating Late Payment will address a specific pain point for businesses.

2.14 Discussion and Conclusion

The European Union's approach to digital legislation is comprehensive, covering various aspects of the digital world. It aims to create a digital ecosystem that promotes innovation, protects fundamental rights, and ensures a secure online environment for all. It is also of highest importance to ensure the coherence of the legislation

Table 2.22 Existing legislations on finance

Legislation (published in the OJEU)	Summary
Common VAT system, (EC) 2006/112, 2022/0407(CNS)	Establishes a harmonized system of value added tax (VAT) across the EU. (Proposal for revision included)
(EC) 2006/112 is available at https://eur-lex.europa.eu/legal-content/EN/TXT/?uri=CELEX%3A02006L0112-20220701 Information on 2022/0407(CNS) is available at https://oeil.secure.europarl.europa.eu/oeil/popups/ficheprocedure.do?reference=2022/0407(CNS)&l=en	
Administrative cooperation in the field of taxation, (EU) 2011/16	This Directive provides the framework for Member States to collaborate and exchange tax information necessary for effective tax administration and enforcement
(EU) 2011/16 is available at https://eur-lex.europa.eu/legal-content/EN/TXT/?uri=CELEX%3A02011L0016-20230101	
Payment Service Directive 2 (PSD2), (EU) 2015/2366 2023/0209(COD)	Regulates payment services (e.g., bank transfers, online payments) within the EU. (Proposal for revision included)
(EU) 2015/2366 is available at https://eur-lex.europa.eu/legal-content/EN/TXT/?uri=CELEX%3A02015L2366-20151223&qid=1688484488618 Information on 2023/0209(COD) is available at https://oeil.secure.europarl.europa.eu/oeil/popups/ficheprocedure.do?reference=2023/0209(COD)&l=en	
Digital Operational Resilience Act (DORA Regulation), (EU) 2022/2554	Establishes a framework for managing operational resilience in the financial sector to ensure critical functions can withstand disruptions
(EU) 2022/2554 is available at https://eur-lex.europa.eu/legal-content/EN/TXT/?uri=CELEX%3A32022R2554&qid=1688484532766	
Crypto-assets Regulation (MiCA), (EU) 2023/1114	Regulates the issuance and trading of crypto-assets (excluding those already covered by existing financial services legislation)
(EU) 2023/1114 is available at https://eur-lex.europa.eu/legal-content/EN/TXT/?uri=CELEX%3A32023R1114&qid=1688484570265	

framework. This highly contributes to its applicability by each European stakeholder as illustrated in Fig. 2.2.

While the present chapter provides a comprehensive overview to the available Digital Legislations to be best knowledge of the authors, subsequent chapters of the present book include a deep-dive into selected initiatives of key importance to stakeholders including industry, small and medium-sized enterprises, academia and others.

There are indeed a number of current initiatives, for example in the field of Cybersecurity, Artificial Intelligence and others which will substantially impact market access requirements to the European Union Single Market. Manufacturers will need to carefully consider those to avoid losing access to this market. In this context, the only legally binding text is the one provided in the documents formally published in the European Union Official Journal—including Legislations, Harmonised Standards developed by the European Standardisation Organisations (ESOs) and similar.

Table 2.23 Planned/ongoing initiatives on finance

Planned/ongoing initiatives	Summary
Financial Data Access Regulation, 2023/0205 (COD)	Aims to improve access to financial data for authorized third-party providers, fostering innovation in financial services
Information on 2023/0205 (COD) is available at https://oeil.secure.europarl.europa.eu/oeil/popups/ficheprocedure.do?reference=2023/0205(COD)&l=en	
Payment Services Regulation, 2023/0210(COD)	Aims to modernize and consolidate existing regulations on payment services (replacing PSD2)
Information on 2023/0210(COD) is available at https://oeil.secure.europarl.europa.eu/oeil/popups/ficheprocedure.do?reference=2023/0210(COD)&l=en	
Digital euro, 2023/0212 (COD)	Proposes the creation of a central bank digital currency (CBDC) issued by the European Central Bank (ECB)
Information on 2023/0212 (COD) is available at https://oeil.secure.europarl.europa.eu/oeil/popups/ficheprocedure.do?reference=2023/0212(COD)&l=en	
Regulation on combating late payment, 2023/0323(COD)	The purpose is to combat late payment in commercial transactions, in order to ensure the proper functioning of the internal market
Information on 2023/0323(COD) is available at https://oeil.secure.europarl.europa.eu/oeil/popups/ficheprocedure.do?reference=2023/0323(COD)&l=en	

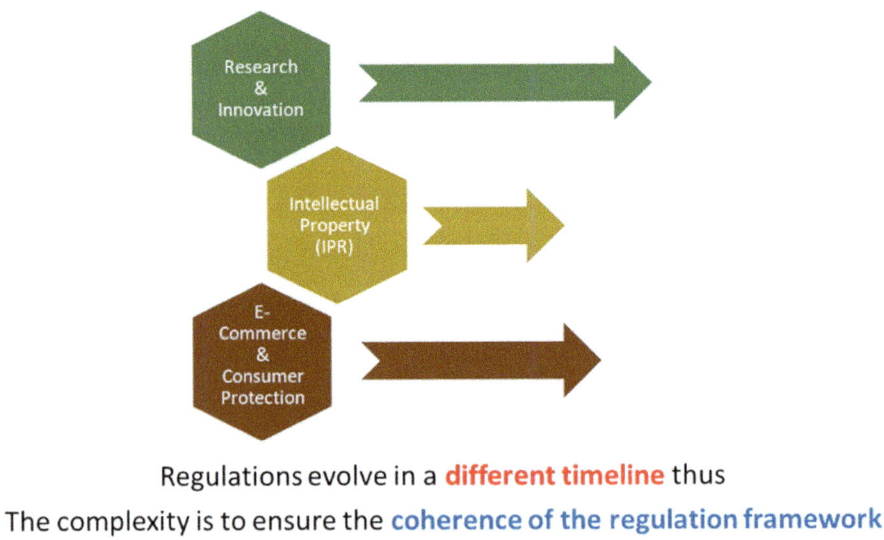

Fig. 2.2 Complexity to ensure the coherence of the legislation framework

Finally, there are still challenges that remain to ensure effective enforcement of legislations across member states, to adapt rapidly to the changes in a fully digital world, and to adjust continuously the legislations to establish the right balance between fostering innovation and ensuring fair competition. For this reason, the legislations presented in this chapter may be revised as appropriate in the future, may be complemented by further initiatives and may be withdrawn as decided by the European Commission, the European Parliament and the European Council.

Disclaimer Note that the authors have assembled the information above to the best of their knowledge as available at the time when the present book was written. Any interpretation is the personal opinion of the authors. Also, the legislation framework can change and evolve any time. The only legally binding text is the one provided in the documents formally published in the European Union Official Journal.

References

1. EUR-Lex, Access to European Union Law, European Commission, service available at https://eur-lex.europa.eu/advanced-search-form.html
2. Newman, A. L. "Digital Policy-Making in the European Union: Building the New Economy of an Information Society." In Policy-Making in the European Union. 8th ed, edited by Helen Wallace, Mark A. Pollack, Christilla Roederer-Rynning, and Alasdair R. Young. Oxford: Oxford University Press, 2020. Politics Trove, 2021. https://doi.org/10.1093/hepl/9780198807605.003.0012.
3. Bruegel, European Think Tank, https://www.bruegel.org/bruegel-european-think-tank-specialises-economics
4. A dataset on EU legislation for the digital world, 16 November 2023, available at https://www.bruegel.org/dataset/dataset-eu-legislation-digital-world
5. Giovanni De Gregorio, The rise of digital constitutionalism in the European Union, International Journal of Constitutional Law, Volume 19, Issue 1, January 2021, Pages 41–70, https://doi.org/10.1093/icon/moab001
6. Quintais, João Pedro, The New Copyright in the Digital Single Market Directive: A Critical Look (October 14, 2019). European Intellectual Property Review 2020(1) (Forthcoming), Available at SSRN: https://ssrn.com/abstract=3424770 or https://doi.org/10.2139/ssrn.3424770
7. Gaie, C., Karpiuk, M. (2024). The Provision of e-Services by Public Administration Bodies and Their Cybersecurity. In: Gaie, C., Mehta, M. (eds) Transforming Public Services—Combining Data and Algorithms to Fulfil Citizen's Expectations. Intelligent Systems Reference Library, vol 252. Springer, Cham. https://doi.org/10.1007/978-3-031-55575-6_7
8. Gaie, C. (2023). Struggling Against Tax Fraud, a Holistic Approach Using Artificial Intelligence. In: Gaie, C., Mehta, M. (eds) Recent Advances in Data and Algorithms for e-Government. Artificial Intelligence-Enhanced Software and Systems Engineering, vol 5. Springer, Cham. https://doi.org/10.1007/978-3-031-22408-9_4

Chapter 3
Introduction to the European Artificial Intelligence Act

Markus Mueck, Christophe Gaie, and Dimitris C. Gkikas

Abstract In July 2024, the European Union Artificial Intelligence Act (EU AI Act) was published in the Official Journal of the European Union. It is a major regulation governing the access to the European Union Single Market for Artificial Intelligence (AI) systems. It is expected that the principles of the EU AI Act will have far reaching consequences beyond the borders of Europe and will influence related debates worldwide. In this chapter, the authors will give an overview of the EU AI Act, including a summary of related institutions which take active roles in the implementation and enforcement, and will discuss possible solutions for meeting inherent requirements. Also, the definitions of so-called High Risk AI Systems and General Purpose AI Models with systemic risk are introduced and discussed; most of the requirements of the AI Act target such categories. Finally, exemptions for scientific research and development and for AI systems released under free and open-source licenses will be discussed.

Keywords Regulation · Digital · Artificial intelligence · Machine learning · European Union Single Market

3.1 Introduction

Following the publication of a first proposal on an EU AI Act [1] in April 2021, and as discussed in [2], the final EU AI Act [3] was published in the EU Official Journal in July 2024 including substantial changes over the initial version while keeping

M. Mueck (✉)
Intel Deutschland GmbH, Neubiberg, Germany
e-mail: Markus.Dominik.Mueck@intel.com

C. Gaie
Head of the Engineering and Digital Innovation Division, Prime Minister Services, Paris, France
e-mail: christophe.gaie@gmail.com

D. C. Gkikas
University of Patras, Patras, Greece
e-mail: dgkikas@upatras.gr

M. Mueck and C. Gaie (eds.), *European Digital Regulations*, Intelligent Systems Reference Library 265, https://doi.org/10.1007/978-3-031-80809-8_3

the overall spirit of the original proposal. The EU AI Act specifically applies to AI systems which are defined as follows by the EU AI Act:

> 'AI system' means a machine-based system that is designed to operate with varying levels of autonomy and that may exhibit adaptiveness after deployment, and that, for explicit or implicit objectives, infers, from the input it receives, how to generate outputs such as predictions, content, recommendations, or decisions that can influence physical or virtual environments.

As outlined in [3], the purpose of the EU AI Act is *to improve the functioning of the internal market by laying down a uniform legal framework in particular for the development, the placing on the market, the putting into service and the use of artificial intelligence systems (AI systems) in the Union, in accordance with Union values, to promote the uptake of human centric and trustworthy artificial intelligence (AI) while ensuring a high level of protection of health, safety, fundamental rights as enshrined in the Charter of Fundamental Rights of the European Union (the 'Charter'), including democracy, the rule of law and environmental protection, to protect against the harmful effects of AI systems in the Union, and to support innovation.*

Finally, as further outlined by a Webinar of the European AI Office [4], the EU AI Act relies on the following main principles, illustrated in Fig. 3.1:

1. Enhanced Product Regulation: addressing risks to health, safety and fundamental rights;
2. Prioritizes the identification of High Risk AI systems and establishes requirements to mitigate the potential risks these systems may generate;
3. Risk-based and lifecycle approach: regulate according to risk, pre- and post-market monitoring;
4. Trust across entire value chain: rules for AI systems and General-Purpose AI (GPAI) models;
5. Responsible innovation: encourage the development of trustworthy and human-centric AI.

Following the publication of the EU AI Act in the European Union Official Journal, all stakeholders need to engage in its implementation.

An overview of the key involved parties is summarized in Fig. 3.2.

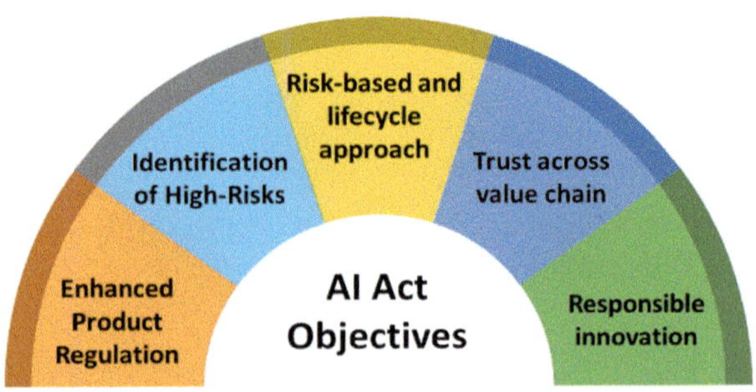

Fig. 3.1 Illustration of the main objectives of the AI Act

Fig. 3.2 Various roles of the key stakeholders

In the following sections of the present chapter, an overview on key parts of the EU AI Act will be given; this includes a summary on the implementation time-line, requirements introduced onto High Risk AI Systems and General Purpose AI (GPAI) Models with systemic risk, a discussion on exemptions such as research and certain open-source solutions, etc. Also, possible implementation solutions are being discussed with the objective to meet the requirements outlined by the EU AI Act.

3.2 Implementation Time-Line

In support of the EU AI Act, European Standardisations Organisations (ESOs) receive Standardisation Requests (SRs) with the task to develop so-called Harmonised European Norms (HENs)—or Harmonised Standards (HS)—or other deliverables. As explained in the Blue Guide on product safety of the European Commission [5], HENs provide presumption of conformity and thus stakeholders can use them to demonstrate that products, services, or processes comply with relevant EU legislation such as the EU AI Act.

In the context of the EU AI Act, a first SR [6] was issued in May 2023 even before the EU AI Act was officially listed in the EU Official Journal. For this reason, only deliverables other than HENs were requested; still, this approach helped to accelerate the overall process.

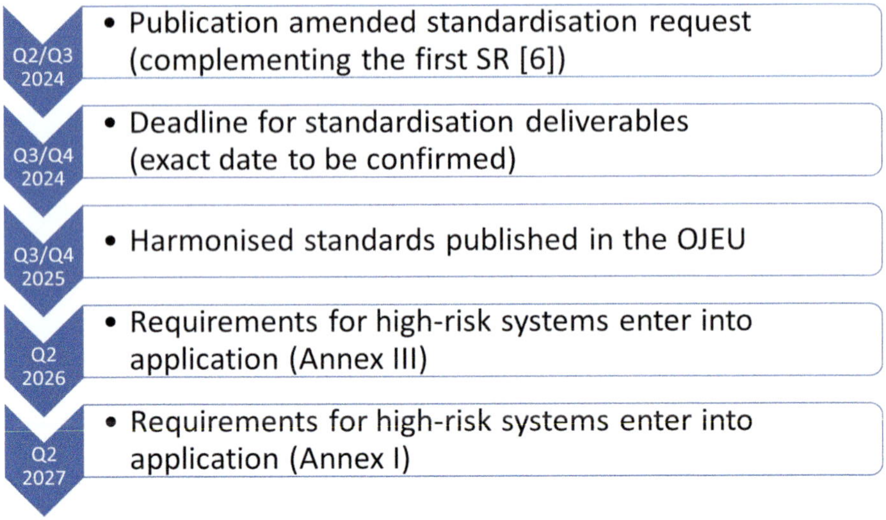

Fig. 3.3 Next steps supporting the implementation of the EU AI Act

ESOs are expected to build on available international standards developed by ISO/IEC. If appropriate, those can indeed be transposed into HENs using the so-called Vienna [7] and Frankfurt [8] agreements between ISO/IEC and the European Committee for Standardization (CEN) and European Committee for Electrotechnical Standardization (CENELEC). Key candidates include ISO/IEC 42001 [9] which introduces a management system for AI.

Subsequent steps are being outlined by a presentation of the European Commission to the ETSI AI Conference in February 2024 [10] as illustrated in Fig. 3.3. Standards related deadlines are initial estimates and may be adapted as needed.

The application of requirements introduced by the EU AI Act will follow a gradual approach as defined in Article 113 of the EU AI Act and illustrated in Fig. 3.4.

Article 113 of the EU AI Act defines the exact dates as follows (Table 3.1).

3.3 Institutions Supporting the Implementation and Enforcement of the EU AI Act

The EU AI Act mentions several institutions which will support its implementation and enforcement. As introduced in [10], the EU AI Act refers specifically to

- **National competent authorities** to supervise High Risk conformity and to enforce market surveillance for Annex I (list of applicable EU harmonisation legislation) and Annex III (list of High Risk AI systems) and the European Data Protection Supervisor (EDPS) for union entities;

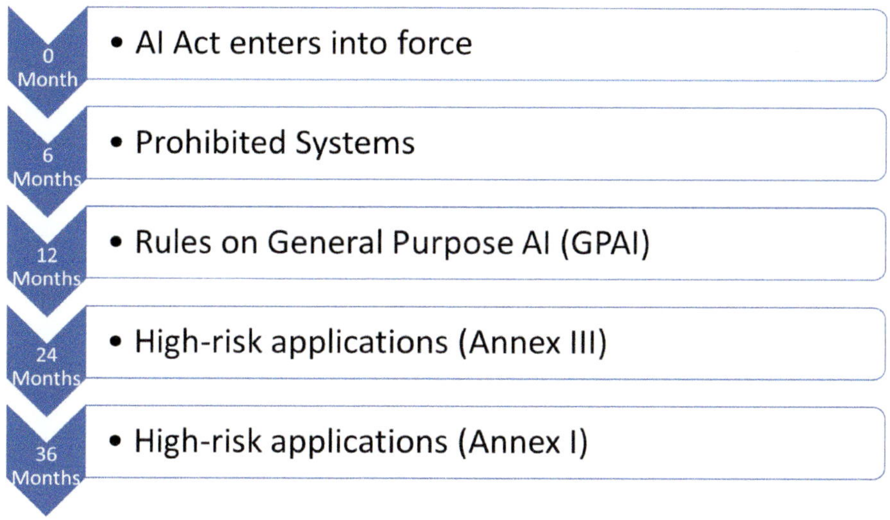

Fig. 3.4 A gradual approach to the application of the EU AI Act

Table 3.1 Dates of applicability of the EU AI Act

Item	Applicable from date on
Regulation as a whole (with the exceptions listed below)	2 August 2026
Chapter I (General Provisions) and Chapter II (Prohibited AI Practices)	2 February 2025
Chapter III (High Risk AI Systems) Section 4 (Notifying authorities and notified bodies), Chapter V (General Purpose AI Models), Chapter VII (Governance) and Chapter XII (Penalties) and Article 78 (Confidentiality)	2 August 2025 with the exception of Article 101 (Fines for providers of general-purpose AI models)
Article 6(1) on classification rules for high-risk AI systems and the corresponding obligations in this Regulation	2 August 2027

- Newly established **AI Office** to supervise general purpose AI models to conduct evaluations, request measures and issue fines; Commission's internal implementation body within the Directorate General for Communications Networks, Content and Technology (DG CNECT);
- **European Artificial Intelligence Board** consists of high-level representatives of competent national supervisory authorities; advising and assisting the Commission (including the AI Office);
- **Advisory Forum** consists of a balanced selection of stakeholders, including industry, start-ups, SMEs, civil society and academia;

- **Scientific Panel** supports the implementation and enforcement of the Regulation as regards GPAI models.

Especially the establishment of the AI Office is noteworthy, as it will be *the center of AI expertise across the EU. It will play a key role in implementing the AI Act—especially for general-purpose AI—foster the development and use of trustworthy AI, and international cooperation* [11].

3.4 A Risk-Based Approach to Regulating AI

The EU AI Act applies a risk based approach to AI—with the term risk to be understood in relation to health, safety and fundamental rights. We note that there exist other definitions, for example Wikipedia relates risk very generally to *uncertainty about the effects/implications* [12] and the ISO 9001 specification defines risk as *the effect of uncertainty on an expected result* [13]. In the context of the AI Act, we thus apply a very specific and narrow meaning of risk. For AI systems, the following 4-step classification approach is applied as outlined in [4] and illustrated in Fig. 3.5.

Article 6 of the EU AI Act defines which AI systems are considered to be "High Risk". Article 6(2) specifically refers to Annex III of the EU AI Act which contains an extensive list of such High Risk AI systems. Those include AI systems for Biometric identification and categorisation of natural persons, for Management and operation of critical infrastructure, etc.

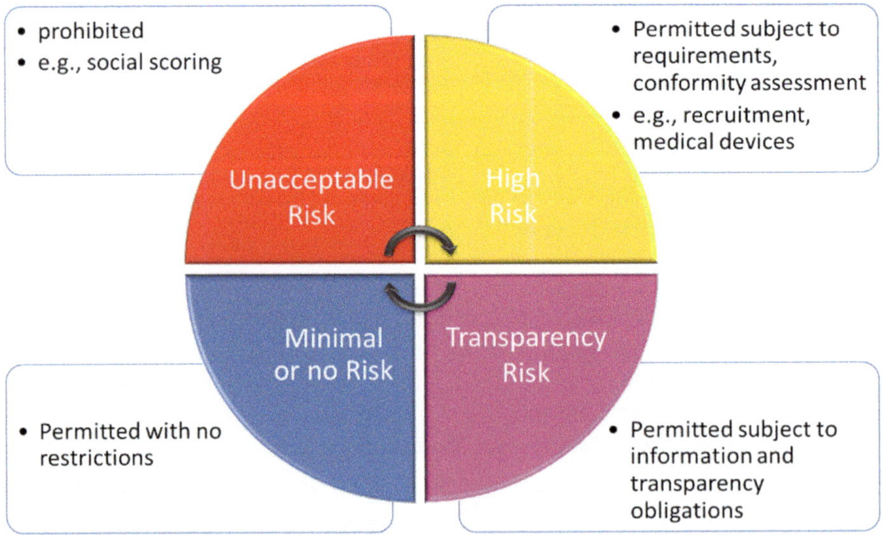

Fig. 3.5 Risk based approach to regulating AI systems

For General Purpose AI (GPAI) Models, the following applies as outlined in [4]:

- For all AI Models: Transparency rules;
- In case of systemic risk: a Risk Management based approach.

Note that systemic risk is defined as follows in the AI Act:

> 'systemic risk' means a risk that is specific to the high-impact capabilities of general-purpose AI models, having a significant impact on the Union market due to their reach, or due to actual or reasonably foreseeable negative effects on public health, safety, public security, fundamental rights, or the society as a whole, that can be propagated at scale across the value chain.

Further details on the classification of general-purpose AI models as general-purpose AI models with systemic risk are outlined in Article 51 of the EU AI Act.

3.5 Discussion of Requirements of the EU AI Act for High Risk AI Systems and General-Purpose AI Models

In the view of the authors of the present chapter, the structure of the EU AI Act can be understood as introducing two main pillars of market access requirements, as illustrated in Fig. 3.6.

Major parts of the EU AU Act specifically focus on High Risk AI Systems and General Purpose AI (GPAI) Models with systemic risk.

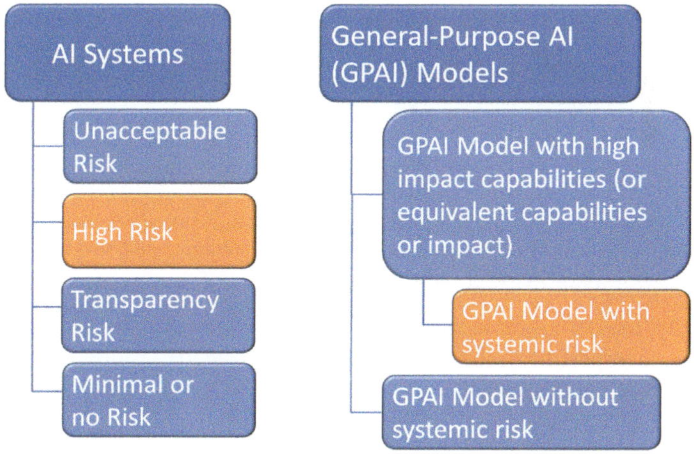

Fig. 3.6 Main pillars of the EU AI Act including High Risk AI systems and general-purpose AI (GPAI) models

3.5.1 High Risk AI Systems

Chapter III of the EU AI Act details how to determine whether an AI System is considered High Risk.

Chapter III
High Risk AI Systems
Section 1
Classification of AI systems as High Risk
Article 6
Classification rules for High Risk AI systems

1. Irrespective of whether an AI system is placed on the market or put into service independently of the products referred to in points (a) and (b), that AI system shall be considered to be High Risk where both of the following conditions are fulfilled:
   (a) the AI system is intended to be used as a safety component of a product, or the AI system is itself a product, covered by the Union harmonisation legislation listed in Annex I;
   (b) the product whose safety component pursuant to point (a) is the AI system, or the AI system itself as a product, is required to undergo a third-party conformity assessment, with a view to the placing on the market or the putting into service of that product pursuant to the Union harmonisation legislation listed in Annex I
2. In addition to the High Risk AI systems referred to in paragraph 1, AI systems referred to in Annex III shall be considered to be High Risk

Furthermore, Article 6(3) introduces a number of derogations for AI systems covered by Article 6(2).

Concerning conformity with essential requirements, Article 40 states that *High Risk AI systems or general-purpose AI models which are in conformity with harmonised standards or parts thereof the references of which have been published in the Official Journal of the European Union in accordance with Regulation (EU) No 1025/2012 shall be presumed to be in conformity with the requirements...* Additional details on conformity assessment for High Risk AI systems are given in Article 43.

In the upper definition of High Risk AI systems, the important notion of a *safety component of a product* is introduced being defined as follows:

> 'safety component' means a component of a product or of an AI system which fulfils a safety function for that product or AI system, or the failure or malfunctioning of which endangers the health and safety of persons or property.

Obviously, the definition is rather generic and requires further clarification. This clarification is especially relevant in light of having the following AI Systems declared as High Risk in Annex III (list of High Risk AI systems) of the EU AI Act:

> Critical infrastructure: AI systems intended to be used as safety components in the management and operation of critical digital infrastructure, road traffic, or in the supply of water, gas, heating or electricity.

For example, in the view of the authors of the present chapter, it is not entirely clear which parts of a cellular communication network (considered to be critical infrastructure according to the Regulation on Resilience of Critical Entities [14]) may or may not fall under such a definition. To provide clarification for Radio Equipment in general, the EC has submitted a related input paper to the EC Expert Group Radio Equipment where corresponding clarifications are proposed when such AI-based safety components are integrated into radio systems [15].

The EC input paper to EC Expert Group Radio Equipment [15] states that *only the following components are covered by this definition:*

- *Strictly safety components, related to the avoidance of damage to users (Article 3(1)(a) of the Radio Equipment Directive);*
- *Cybersecurity components (Articles 3(3)(d-f) of the Radio Equipment Directive);*
- *Components ensuring access to certain emergency services (Article 3(3)(g) of the Radio Equipment Directive).*

It is furthermore noted by [15] that *components with functionalities linked to other essential requirements of the RED (such as frequency management or electromagnetic compatibility) do not meet this requirement* [First condition: AI-based safety component].

Following the analysis in [15], it is stated that *in conclusion, a radio equipment device qualifies as High Risk AI system if it incorporates an AI-based component related to only cybersecurity (as described in Articles 3(3)(d-f) of the RED) or access to emergency services (as described in Article 3(3)(g) of the RED), irrespectively of the existence and/or application of relevant harmonised standards in support of the RED.*

It is expected that the related discussion will further evolve. For further information on the Radio Equipment Directive (RED), it is recommended to consult the corresponding chapter in the present book.

On the other hand, however, Recital 55 of the AI Act states that

> … Safety components of critical infrastructure, including critical digital infrastructure, are systems used to directly protect the physical integrity of critical infrastructure or the health and safety of persons and property but which are not necessary in order for the system to function. The failure or malfunctioning of such components might directly lead to risks to the physical integrity of critical infrastructure and thus to risks to health and safety of persons and property. Components intended to be used solely for cybersecurity purposes should not qualify as safety components. Examples of safety components of such critical infrastructure may include systems for monitoring water pressure or fire alarm controlling systems in cloud computing centres.

In the views of the authors of the present chapter, the upper clarifications seem to be contradictory to some extent—especially to which extent components for cybersecurity purposes imply a classification as high risk—and thus further clarification on the scope of a *safety component* would be welcome.

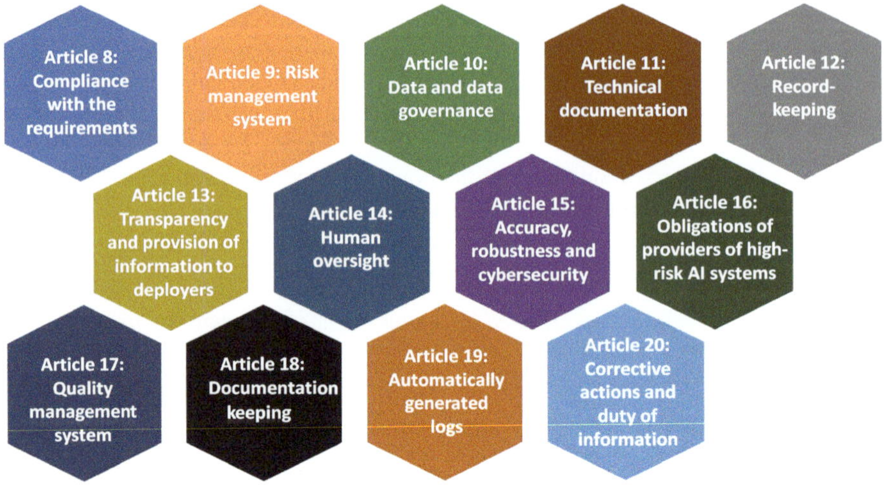

Fig. 3.7 Summary of technical articles of the EU AI Act [3]

Concerning High Risk AI Systems, the EU AI Act sets out requirements manufacturers must follow to comply with the Act. A simplified summary of the corresponding Articles under Chapter III, Section 2 (Articles 8 to Article 15 on ***Requirements for high-risk AI systems***) and Section 3 (Article 16 to Article 27 on ***Obligations of providers and deployers of high-risk AI systems and other parties***) is illustrated in Fig. 3.7 and discussed in the following.

Article 8: Compliance with the requirements spells out the obgligation for High Risk AI systems to comply with the requirements laid out in Section 2 of the EU AI Act on requirements for High Risk AI systems. For this purpose, the intended purpose of the concerned AI systems as well as the generally acknowledged state of the art on AI and AI-related technologies need to be considered.

Article 9: Risk management system introduces the requirement to establish, implement, document and maintain an appropriate risk management system. The corresponding process will be maintained iteratively over the lifetime of the AI system. Furthermore, test requirements are introduced to be applied to concerned AI system with the objective to identify the most appropriate and targeted risk management measures.

Article 10 Data and data governance details requirements on training, validation and testing data sets. Specific attention is drawn to the occurrence of (undesired) biases have the potential to (negatively) affect the health and safety of persons, have a negative impact on fundamental rights or lead to discrimination prohibited under Union law. Furthermore, requirements are introduced on the concerned training, validation and testing data sets to have appropriate statistical properties, to be relevant, sufficiently representative, and to the best extent possible, free of errors and complete in view of the intended purpose.

The authors of the present chapter note that biases are not necessarily undesired or have a negative impact in all cases. To give an example, a breast cancer detection system is typically biased towards women because this is the application base. It is thus the authors' view that it is not appropriate to strictly avoid all types of biases but only those leading to negative effects.

Article 11 Technical documentation details requirements on the technical documentation of the AI system, especially to provide proof that the High Risk AI system complies with the requirements set out and to give relevant information to national competent authorities and notified bodies. This documentation needs to be made available prior to placement of the AI system onto the European Union Single Market. To support Small and Medium-sized Enterprises (SMEs) including start-ups, a simplified process is introduced for such stakeholders.

Article 12 Record-keeping instructs manufacturers of AI systems to automatically record relevant events (or logs) over the lifetime of the system which ensures a level of traceability of the functioning of a High Risk AI system.

Article 13 Transparency and provision of information to deployers requires that the operation of the concerned AI system is sufficiently transparent and deployers are able to interpret a system's output and use it appropriately. This includes the need for the manufacturer to provide concise, complete, correct and clear instructions. Minimum information to be included is detailed in the Article.

Article 14 Human oversight introduces a need to have appropriate human–machine interface tools implemented which are able to interact with a natural person overseeing the operation of the AI system. As it is further stated in the Article, *human oversight shall aim to prevent or minimise the risks to health, safety or fundamental rights that may emerge when a High Risk AI system is used in accordance with its intended purpose, and under conditions of reasonably foreseeable misuse.* The human supervisor must be able to take action, including to disregard, override or reverse the output of the High Risk AI system. A "stop button" is introduced as a suitable means for triggering an emergency interruption of the AI system.

Article 15 Accuracy, robustness and cybersecurity further details requirements for the design and development phase of an AI system, including the need to offer an appropriate level of accuracy, robustness, and cybersecurity. Those characteristics must be maintained over the life-time of the AI system. In terms of robustness, the article states that *the robustness of High Risk AI systems may be achieved through technical redundancy solutions, which may include backup or fail-safe plans.* Finally, requirements are introduced related to the potential development of (undesired) biases during on-going retraining of the AI system over its lifetime and related to being resilient against attempts by unauthorised third parties.

Article 16 Obligations of providers of High Risk AI systems introduces requirements for providers of High Risk AI systems, including the need to meet all requirements of Section 2 of the EU AI Act (Article 8 to Article 15 as discussed above), to provide required information (contact address, etc.), to demonstrate the availability of automated logging, to affix the CE marking, etc. Before placement onto the European Union Single Market, the High Risk AI system must undergoes the relevant conformity assessment procedure.

Article 17 Quality management system furthermore requires providers of High Risk AI systems to implement a quality management system that ensures compliance with the EU AI Act—be documented in a systematic way, including information such as applied technical specifications, including standards, as well as procedures related to examination, test and validation to be carried out before, during and after the development of the High Risk AI system, etc. Specific instructions are given to providers that are financial institutions.

Article 18 Documentation keeping summarizes what the provider of a High Risk AI system must maintain and provide to national competent authorities for up to10 years after the High Risk AI system has been placed on the market or put into service. This includes the technical documentation, documentation on the risk management system, etc.

Article 19 Automatically generated logs details the maintenance of logs, requiring them to be kept *for a period appropriate to the intended purpose of the High Risk AI system, of at least six months, unless provided otherwise in the applicable Union or national law, in particular in Union law on the protection of personal data.*

Article 20 Corrective actions and duty of information introduces steps to be taken ("corrective measures") by providers of AI system in case that they do not operate in conformity of the EU AI Act and they have already been places onto the European Union Single Market or are in service.

Chapter III, Section 3 of the EU AI Act furthermore introduces more generic obligations on the Cooperation with competent authorities (Article 21), the designation of authorised representatives of providers of High Risk AI systems (Article 22), obligations of importers (Article 23), obligations of distributors (Article 24), responsibilities along the AI value chain (Article 25), obligations of deployers of High Risk AI systems (Article 26) and fundamental rights impact assessment for High Risk AI systems (Article 27).

3.5.2 General Purpose AI Models (GPAI Models)

Concerning General Purpose AI Models, Chapter V, Section 2, Article 53 of the EU AI Act introduces related *obligations for providers of general-purpose AI models.* Those obligations specifically refer to appropriate documentation such as the requirement to *draw up and keep up-to-date the technical documentation of the model, including its training and testing process and the results of its evaluation.* It is noteworthy that the Section also refers to codes of practice as follows.

Providers of general-purpose AI models may rely on codes of practice within the meaning of Article 56 to demonstrate compliance with the obligations set out in paragraph 1 of this Article, until a harmonised standard is published.

At the time when the present chapter was written, such codes of practice were not yet available. They are expected to be developed in the sequel of the implementation process of the EU AI Act according to its Article 56.

Whether or not a General-Purpose AI Model is considered of *systemic risk* is laid out in Article 51 of the EU AI Act with additional information in Annex XIII. For that purpose, to mention one of several provisions in the EU AI Act, the Model needs to be assessed whether it *has high impact capabilities evaluated on the basis of appropriate technical tools and methodologies, including indicators and benchmarks.* More specifically, a General-Purpose AI model *shall be presumed to have high impact capabilities … when the cumulative amount of computation used for its training measured in floating point operations is greater than 10^25.*

Section 3 of the EU AI Act furthermore details *obligations of providers of general-purpose AI models with systemic risk.* Those include a requirement *to perform model evaluation in accordance with standardised protocols and tools reflecting the state of the art, to assess and mitigate possible systemic risks, to ensure an adequate level of cybersecurity protection* and others.

3.6 Standards in Support of the EU AI Act

With the EU AI Act being a regulation under the New Legislative Framework (NLF), European Standardisation Organisations (ESOs) are being tasked to develop supporting standards, including Harmonised European Norms which provide presumption of conformity with the regulation and allow for self-declaration of conformity by the concerned stakeholders.

As outlined by a presentation of the European Commission at the ETSI AI Conference 2024 [10] illustrated in Fig. 3.8, the standardisation process will be executed in multiple steps over the 2020–2027 time period (and possibly beyond).

The initial Standardisation Request [6] supporting the implementation of the EU AI Act was issued to CEN and CENELEC while requiring to establish a close relationship with ETSI such that all European Standardisation Organisations are involved:

> (16) Given that the European Telecommunications Standards Institute (ETSI) has already done work and possesses specific expertise on certain matters relevant to the request, such as security, it is appropriate that European standards or European standardisation deliverables drafted on the basis of this request take into account, as appropriate, the relevant ETSI's work.

The Joint Research Center of the European Commission (JRC) has studied available standards of potential relevance to the implementation of the EU AI Act. The final report [16] presents an extensive summary referencing standards developed by ISO/IEC, IEEE, ETSI and ITU-T, all mapped onto the key areas of the EU AI Act:

- Data and data governance;
- Risk management system;
- Technical data and Record keeping;
- Transparency and information to users;
- Human oversight;

Fig. 3.8 Sequence of actions to implement EU AI Act [10] (time lines updated according to the published EU AI Act)

- Accuracy, robustness, and cybersecurity;
- Quality management system.

Following the acceptance of the Standardisation Request [6], CEN/CENELEC developed a Standards Architecture which was presented at the ETSI AI Conference in a presentation by CEN/CENELEC [17] and displayed in Fig. 3.9.

CEN/CENELEC has agreements in place with the International Organization for Standardization (ISO) and International Electrotechnical Commission (IEC), the so-called Vienna [7] and Frankfurt [8] agreements. Those agreements allow for usage of

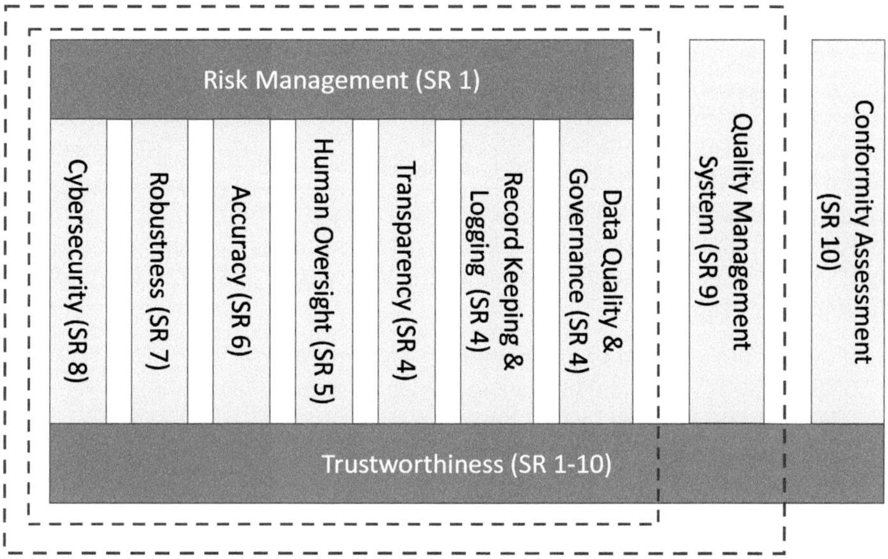

Fig. 3.9 Standards Architecture in support of the EU AI Act by CEN/CENELEC [17]

International Standards developed by ISO/IEC when meeting European legislative and market requirements.

It is indeed common practice to publish specific ISO/IEC specifications as Harmonised European Norms to support the implementation of related European Policy priorities. In the context of the EU AI Act, ISO/IEC has a substantial collection of available standards to offer which have been mainly developed by the ISO/IEC JTC1 SC42, which is a standardization subcommittee of the Joint Technical Committee ISO/IEC JTC 1 of the International Organization for Standardization and the International Electrotechnical Commission.

Examples of standards developed by ISO/IEC, under processing or already published by CEN/CENELEC and of potential relevance to the implementation of the EU AI Act are as follows:

- EN ISO/IEC 22989:2023: Artificial intelligence concepts and terminology (ISO/IEC 22989:2022)
- EN ISO/IEC 23053:2023: Framework for Artificial Intelligence (AI) Systems Using Machine Learning (ML) (ISO/IEC 23053:2022)
- EN ISO/IEC 42001: Information technology—Artificial intelligence—Management system
- EN ISO/IEC 23894: Guidance on risk management
- EN ISO/IEC 24029-2: Assessment of the robustness of neural networks—Part 2: Methodology for the use of formal methods
- EN ISO/IEC 24029-3: Assessment of the robustness of neural networks—Part 3: Methodology for the use of statistical methods.

While most Standards Developing Organizations (SDOs) have dedicated committees developing Artificial Intelligence related standards, ETSI implements a more distributed approach. The field of Artificial Intelligence is large, so that a variety of distinct competences are required, each tailored to the specific sub-fields such as Cybersecurity, Explainability, Trustworthiness, etc. Still, much of ETSI's work in the field of Artificial Intelligence is homed in the Technical Committee Securing Artificial Intelligence (TC SAI). As summarized in the presentation by the Chair of TC SAI at the ETSI AI Conference in February 2024, examples of standards developed by ETSI and of potential relevance for the implementation of the EU AI act include the following [18]:

- DTS/SAI-005 (TS 104 050) AI Ontology and definitions
- DTR/SAI-007 (TR 104 048) Data supply chain report
- DTR/SAI-009 (TR) Mitigation strategy report
- DTR/SAI-008 (TR) Problem statement
- DTS/SAI-006 (TS 104 033) AI Computing Platform Security Framework
- DTR/SAI-004 (TR 104 032) Traceability of AI Models
- DTR/SAI-003 (TR 104 031) Collaborative Artificial Intelligence
- DTR/SAI-0011 (TR) Security aspects of using AI/ML techniques in telecom sector
- DTR/SAI-0010 (TR) Automated Manipulation of Multimedia Identity Representations
- DTR/SAI-001 (TR 104 029) Global AI Security Ecosystem
- DTR/SAI-002 (TR 104 030) AI Critical Security Controls
- DTR/SAI-0013 (TR) AI Act mapping and gap analysis.

A further item under current debate is how to address specific sectorial requirements. Application areas such as Industrial Automation, Health, Aviation, etc. have specific requirements which can deviate substantially from related requirements of the consumer industry for example. It is therefore expected that a combination of Horizontal and Vertical Standards will be developed as illustrated in Fig. 3.10.

A specific vertical industry, such as Industrial Automation, Health, etc. for example, would thus need to meet the Horizontal Standards which are applicable to all application fields together with the applicable Vertical Standards introducing specific requirements for the concerned industry.

To summarize, at the time when the present chapter was written, it was not yet clear which requirements will finally be contained in future Harmonised European Norms (HENs)—or Harmonised Standards (HS)—in support the EU AI Act and providing presumption of conformity. Many of the available standards are high level, introduce requirements on an organization level, etc. It is currently unclear how they will be used to determine whether or not a specific AI system is meeting conformity requirements or not. The initial Standardisation Request [6] indeed states the following:

> The European standards and European standardiztion deliverables shall provide, to the necessary extent and taking into account the state of the art, technology-based, process-based or methodology-based technical specifications in relation to the design and development of

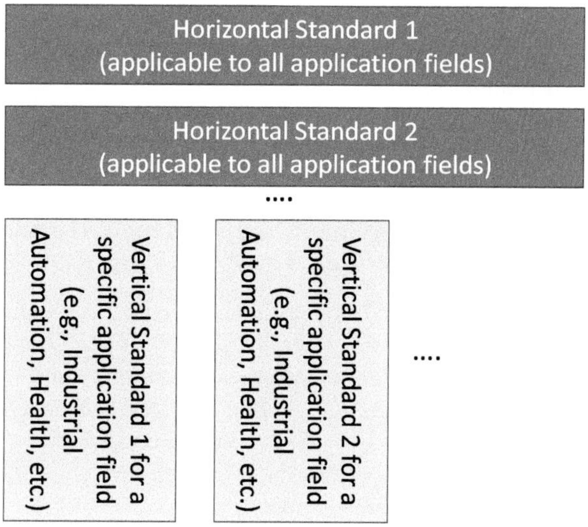

Fig. 3.10 Horizontal and Vertical Standards in support of the EU AI Act

> AI systems, including verification, validation and testing procedures, objectively verifiable criteria and implementable methods to assess compliance with such specifications.

Ideally, there should be quantifiable metrics meeting the call for "*objectively verifiable criteria*". Once those metrics are available, it would be possible to assess whether a given AI System meets the threshold requirements to achieve conformity to the EU AI Act or whether it fails to do so.

3.7 Conformity Assessment

In the context of the EU AI Act, the term *conformity assessment* is defined as follows:

> 'conformity assessment' means the process of demonstrating whether the requirements set out in Chapter III, Section 2 relating to a high-risk AI system have been fulfilled.

Article 42 of the EU AI Act provides details on the *presumption of conformity with certain requirements*. Especially, Article 42(2) relates to the certification of High Risk AI systems which are *certified or for which a statement of conformity has been issued under a cybersecurity scheme pursuant to Regulation (EU) 2019/881* (the Cyber Security Act [19]).

The important role of *Harmonised standards and standardisation deliverables* is outlined in Article 40 of the EU AI Act. It is explained under which conditions Harmonised Standards listed in the EU Official Journal are applicable for presumption of conformity for High Risk AI systems or general-purpose AI models.

Article 42 details the process for *presumption of conformity with certain requirements*. It is important to follow the exact wording of the Article, below some basic principles are summarised in a simplified manner:

- For High Risk AI systems listed in point 1 of Annex III (i.e., *Biometrics*), the following mechanisms apply in combination with Harmonised Standards or Common Specifications:
  - Internal Control (according to Annex VI of the EU AI Act), i.e. self-assessment of conformity by the Manufacturer; or
  - Notified Body based assessment (see Annex VII of the EU AI Act); this approach also needs to be taken if Harmonised Standards or Common Specifications are not available or only partly applied.
- For High Risk AI systems listed in point 2 to 8 of Annex III (i.e., *Critical infrastructure, Education and vocational training, Employment, workers' management and access to self-employment, Access to and enjoyment of essential private services and essential public services and benefits, Law enforcement, Migration, asylum and border control management and Administration of justice and democratic processes*), the following mechanisms apply (noting that they do not provide for the involvement of a Notified Body):
  - Internal Control (according to Annex VI of the EU AI Act), i.e. self-assessment of conformity by the Manufacturer.
- For High Risk AI systems listed in Annex I, Section A of the EU AI Act (legislation under the New Legislative Framework), the relevant conformity assessment procedure applies as required under those legal acts.

3.8 Derogations and Exemptions

The EU AI Act defines its scope in Article 2 and specifically introduces limitations to the application of the regulation. Following those definitions, the EU AI Act does not apply to areas outside the scope of Union law, the EU AI Act does not apply to AI systems used for military, defence or national security purpose, etc. Furthermore, there is a noticeable exemption for research: The EU AI Act *does not apply to AI systems or AI models, including their output, specifically developed and put into service for the sole purpose of scientific research and development*. Also, with some limitations, there is an exemption for open-source software: The EU AI Act *does not apply to AI systems released under free and open-source licences, unless they are placed on the market or put into service as High Risk AI systems or as an AI system that falls under Article 5 (prohibited AI practices) or 50 (transparency obligations for providers and deployers of certain AI systems)*. Article 50 relates, among others, to AI systems intended to interact directly with natural persons, AI systems generating synthetic audio, image, video or text content, emotion recognition system or a biometric categorisation system, etc. Further open-source related exemptions are introduced throughout the EU AI Act, for example Article 25 (Responsibilities along the AI value chain) states

> 4. The provider of a High Risk AI system and the third party that supplies an AI system, tools, services, components, or processes that are used or integrated in a High Risk AI system shall, by written agreement, specify the necessary information, capabilities, technical access and other assistance based on the generally acknowledged state of the art, in order to enable the provider of the High Risk AI system to fully comply with the obligations set out in this Regulation. This paragraph shall not apply to third parties making accessible to the public tools, services, processes, or components, other than general-purpose AI models, under a free and open-source licence.

Within the AI Act, specific care is taken to protect start-ups, Microenterprises as well as small and medium-sized enterprises (SMEs). For this purpose, Article 62 introduces *measures for providers and deployers, in particular SMEs, including start-ups*. Those include priority access to the AI regulatory sandboxes, to raise awareness and to offer training activities on the application of the AI Act, to provide advice and response to queries about the implementation of the AI Act, etc.

Furthermore, Article 63 of the AI Act introduces specific derogations for Microenterprises stating those *may comply with certain elements of the quality management system required by Article 17 of this Regulation in a simplified manner*. Article 17 introduces quality management system requirements which may indeed be challenging to meet for small organizations, for example requirements related to the *design, design control and design verification of the High Risk AI system*. For this purpose, the European Commission will develop corresponding guidelines.

3.9 The AI Liability Directive

At the time when the present chapter was written, a draft of the AI Liability Directive [20] was available which needs to be understood in complement to the EU AI Act. The European Commission has furthermore published material on its background and a corresponding impact assessment [21]. Also, this initiative should be considered in conjunction with the proposed revision of the Product Liability Directive [22].

The purpose of the draft AI Liability Directive [20] is to introduce *non-contractual civil liability rules to artificial intelligence*. It identifies the following challenges in terms of existing liability rules:

> Current national liability rules, in particular based on fault, are not suited to handling liability claims for damage caused by AI-enabled products and services. Under such rules, victims need to prove a wrongful action or omission by a person who caused the damage. The specific characteristics of AI, including complexity, autonomy and opacity (the so-called "black box" effect), may make it difficult or prohibitively expensive for victims to identify the liable person and prove the requirements for a successful liability claim.

To address the above stated challenges and to ensure that persons harmed by artificial intelligence systems enjoy the same level of protection as persons harmed by other technologies, the draft AI Liability Directive provides specific introductions to courts, for example:

> To maintain the balance between the interests of the parties involved in the claim for damages and of third parties concerned, the courts should order the disclosure of evidence only where this is necessary and proportionate for supporting the claim or potential claim for damages…

As discussed by the Tech Law Blog [23], the corresponding rules may have far reaching consequences for AI system providers. In the end, those will potentially be liable for any defective AI systems that are placed on the market.

3.10 Discussion of Implementation Solutions for High Risk AI Systems

Section 2 and Section 3 of the EU AI Act includes Article 9 to Article 20 outlining requirements which demand specific actions by the manufacturer and/or operator of a High Risk AI system.

In order to meet the requirements set out in Article 9 to Article 20, the authors of the present chapter propose the introduction of a number of system entities which are summarised in Fig. 3.11. It should be noted that the proposal is meant to serve as a starting point and any final product is required to pass all conformity assessment steps.

The proposed entities are further detailed below.

The "**Entity for AI Processing**" is the core of the AI System and is typically being trained using some training data set and optionally some additional data that is being acquired while the AI system is being operated. Such an AI system can rely on various machine learning methods/approaches including

- Regression,
- Classification,
- Clustering,
- Dimensionality Reduction,
- Ensemble Methods,
- Neural Nets and Deep Learning,
- Transfer Learning,
- Reinforcement Learning,
- Natural Language Processing,
- Word Embeddings.

The "**Entity for Self-Verification**" is proposed to verify the correct operation of the newly trained system. It is in particular proposed to use a pre-defined test-data set (which is different from the data set used for training) as input to the AI system in order to verify the correct operation. Only if the correct operation is verified, the AI system is allowed for full usage for its intended purpose. In the opposite case, e.g. in case that biases are detected or any unexpected behavior, the operation of the AI system is interrupted until the issues are resolved. Such a verification step is periodically repeated in case of retraining of the AI system with new data.

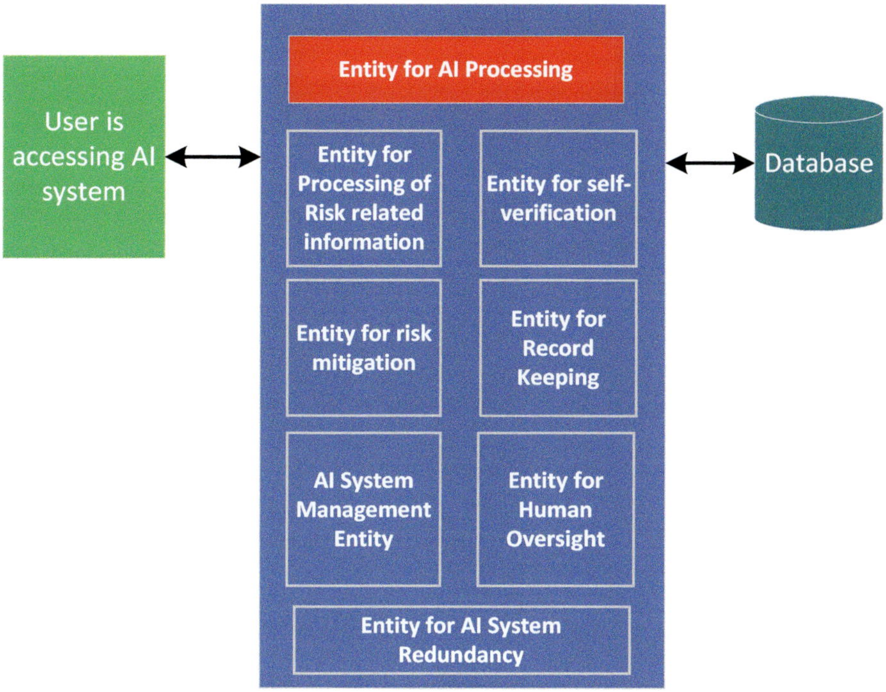

Fig. 3.11 Proposed entities of a High Risk AI system

When the system is finally used for its intended purpose (after all successful verification steps), the "**Entity for Record Keeping**" will be logging all user interactions (commands given by the authorized user, etc.) and will record the behavior of the AI system and store relevant information in the database.

The "**Entity for Risk Mitigation**" will propose a trade-off between risk and functionality to the user. For example, the entity may propose that the system is constantly retrained using the observed information obtained during operation. The upside is that this may improve the quality of the AI decision making. The risk is that the new data may introduce biases or other undesired characteristics. The authorized User will need to decide whether or not corresponding risks are being taken to achieve the expected improvements.

The "**Entity for Processing Risk of related Information**" will take the results of other entities, such as the "Entity for Self-Verification" and process identified risk related information and unexpected behavior information such that it can be presented in a concise way to the authorized user (e.g., by illustrating statistics on the decision making, including outlining of unexpected biases of the statistics, etc.). In case of an issue, the user can use the provided information to take action, e.g. to terminate the AI system operation through the "Entity for Human Oversight".

The “**Entity for Human Oversight**” will allow the authorized user to take action in case that the AI system operates in an unexpected or undesired way, e.g. in case that the decision making processes indicate (undesired) biases. The user may then take several actions, including termination of the AI system operation, enforce a retraining of the system, choose a different risk trade-off through the “Entity for Risk Mitigation”, etc.

The “**Entity for AI System Redundancy**” oversees redundant replacement options for critical entities, components or elements of the AI system. In case that some malfunctioning entity, component or element is identified, typically relying on information by the “Entity for Self-Verification”, then this entity is used to configure a corresponding replacement. After the replacement, the correct operation of the AI system is typically verified, again relying on information by the “Entity for Self-Verification”. If it is successfully verified, then the operation of the AI system may continue.

The “**AI System Management Entity**” will orchestrate the interaction between the different building blocks indicated above. For example, when one of the entities of an AI system is dysfunctional or operates in an unexpected way, then the “AI System Management Entity” may detect this behavior relying on information by the “Entity for Self-Verification” and may trigger the replacement of concerned components/entities by redundant replacement components/entities through the “Entity for AI System Redundancy”.

3.11 Interdependencies with Other European Regulations

In terms of Cybersecurity, the European Artificial Intelligence Act (AI Act) [3] specifically refers to the European Cybersecurity Act, i.e. Regulation (EU) 2019/881 [19]. The AI Act acknowledges interdependencies and states that the *European Parliament and of the Council (37), the Commission should cooperate with ENISA on issues related to cybersecurity of AI systems*. The AI Act furthermore states in *Article 42: Presumption of conformity with certain requirements*:

> High Risk AI systems that have been certified or for which a statement of conformity has been issued under a cybersecurity scheme pursuant to Regulation (EU) 2019/881 and the references of which have been published in the Official Journal of the European Union shall be presumed to comply with the cybersecurity requirements set out in Article 15 of this Regulation in so far as the cybersecurity certificate or statement of conformity or parts thereof cover those requirements.

Article 15 of the AI Act specifically relates to *accuracy, robustness and cybersecurity*.

3.12 Regulation Activities in the Field of Artificial Intelligence in Other Regions

While the initiative of the EU AI Act was certainly one of the earliest regulation efforts in the field of Artificial Intelligence around the world, other regions started to follow and are also undertaking related actions. It is expected, though, that the EU AI Act is substantially influencing the global debate and will be a reference for other policy discussions outside of Europe. In many case, a cross-regional dialogue is implemented as it is for example the case for the EU-US Trade and Technology Council between the EU and United States [24]. In the following, the regulation situation will be discussed for the countries or regions displayed in Fig. 3.12.

3.12.1 United States

In the **United States**, the US White House has published the "Executive Order on the Safe, Secure, and Trustworthy Development and Use of Artificial Intelligence" on 30th October 2023 [25]. The US Executive Order [25] tasks the US National Institute of Standards and Technology (NIST) to develop a number of deliverables. In a recent communication by NIST [26], a correspondig implementation plan is outlined. Specifically, NIST has *released four draft publications intended to help improve the safety, security and trustworthiness of artificial intelligence (AI) systems.* To outline some of the differences with the EU AI Act, we note that the EU AI Act is very focused on High Risk AI Systems and complex General Purpose AI Systems; the US Executive order on the other hand, is specifically concerned with *outputs that may represent nuclear, nonproliferation, biological, chemical, critical infrastructure,*

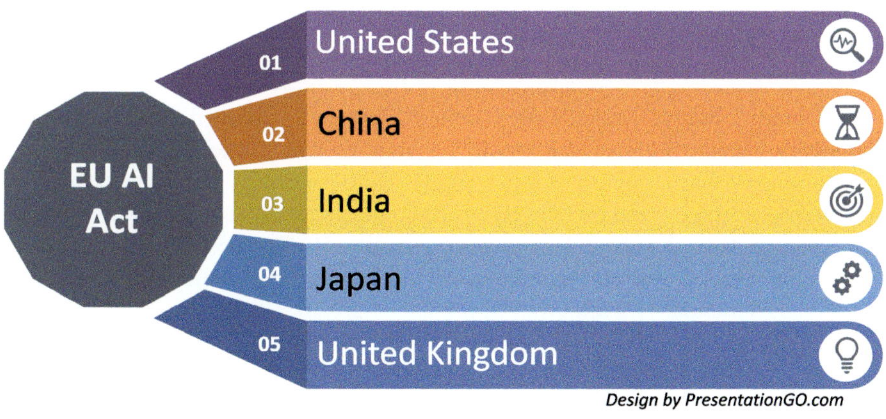

Fig. 3.12 AI related regulation activities are discussed for the United States, China, India, Japan and the United Kingdom

and energy-security threats or hazards. Still, there are also numerous similarities. For example, the US Executive Order introduces reporting requirements for *any model that was trained using a quantity of computing power greater than 10^26 integer or floating-point operations* [25]. The EU AI Act, on the other hand, introduces the notion of AI models of "high impact capabilities" and applies similar conditions: *A general-purpose AI model shall be presumed to have high impact capabilities pursuant to paragraph 1, point (a), when the cumulative amount of computation used for its training measured in floating point operations is greater than 10^25* [3].

Furthermore, the **United States** has authorized additional executive orders including *Maintaining American Leadership in AI* and *Promoting the Use of Trustworthy AI in the Federal Government Act*. Also, some legislations and proposed bills have been activated like *AI Training Act, National AI Initiative Act, Algorithmic Accountability Act (proposed), Transparent Automated Governance Act (proposed), and Global Technology Leadership Act (proposed)* [27].

To obtain further information on other regions beyond Europe and the United States, we rely on presentations given at the ETSI AI Conference, held in February 2024, on the status in China, India, Japan and the UK [28–31].

3.12.2 China

In **China**, there is currently no initiative comparable to the EU AI Act as reported by in [28]. Still, there are policy initiatives related to Artificial Intelligence as follows:

- Personal Information Protection Law of the People's Republic of China;
- China Data Security Law;
- New Generation of AI Ethics Norms (Sep 2021) by Ministry of Science and Technology (MoST):—Guidelines for natural persons, legal persons, and other relevant institutions that are engaged in AI activities;
- Interim Measures for the Administration of Generative Artificial Intelligence Services (Measures).

The upper initiatives include provisions for content generated via AI or related technology. The provisions focus mainly on ensuring that synthetic content is appropriately marked and recorded as required. Measures primarily aim to guarantee the authenticity, accuracy, and objectivity of the content generated by generative AI. On 18th October 2023, the China Cyber Administration CAC issued the China proposal on the "Global AI Governance Initiative" with the objectives detailed in Fig. 3.13.

China's AI Standard Development Organizations can be differentiated as indicated in Fig. 3.14.

Ina nutshell, Chinese Standards are generally national specific, and SAC/TC260/SWG-BDS, the SDO responsible for AI security/safety standards development, rarely adopts or converts the international standards.

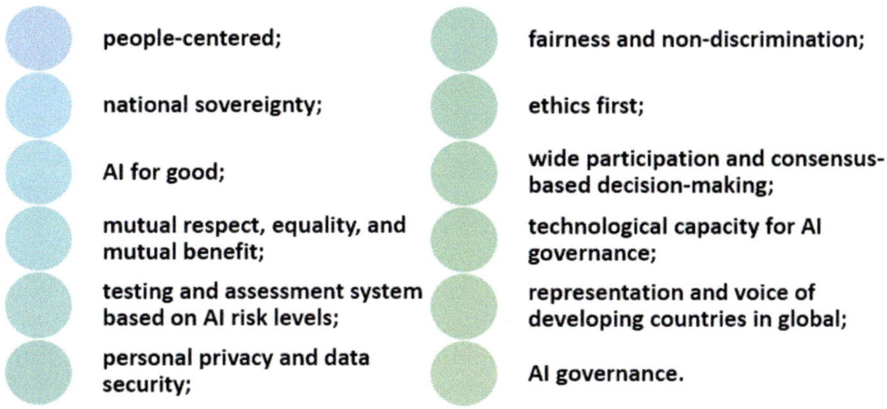

Fig. 3.13 Objectives of the China proposal on the "Global AI Governance Initiative"

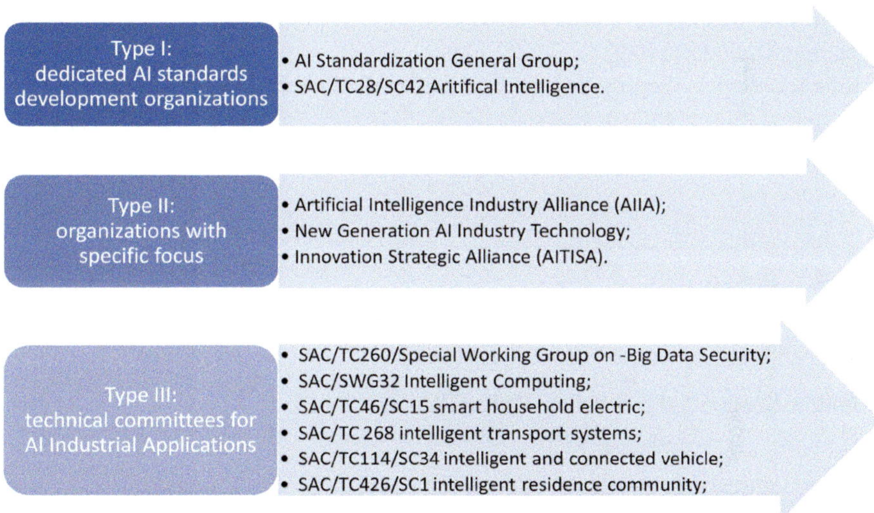

Fig. 3.14 China's AI Standard Development Organizations

3.12.3 *India*

In **India**, there is a stated interest in Artificial Intelligence as outlined in [29]:

- As discussed by an IDC report, the Artificial Intelligence market in India is likely to reach USD 7.8 billion by 2025 from originally USD 3.1 billion in 2020;
- Artificial Intelligence applies to various sectors: smart manufacturing, robots, autonomous cars, virtual reality, healthcare, interactive speech interfaces, visual

recognition, data analysis/manipulation, home appliances, cybersecurity or spatial programming;
- Artificial Intelligence is expected to have a high or a very high impact on business areas that are "entirely unknown to the company today" according to responses by over 60%-70% of companies queried;
- The GPAI Summit was hosted by the Chair of the Global Partnership on Artificial Intelligence (GPAI) in December 2023 in India.

In August 2017, the "Task Force on AI for India's Economic Transformation" was created by India's Ministry of Commerce and Industry. It has outlined the focus areas detailed in Fig. 3.15:

The Ministry of Electronics and Information Technology (MEITY) had created four committees to create a policy framework and to develop the ecosystem for Artificial Intelligence. These committees have released following reports on AI:

- Report on platforms and data on AI;
- Report on leveraging AI for identifying national missions in key sectors;
- Report on mapping technological capabilities, key policy enablers required across sectors, skilling, reskill;
- Report on cyber security, safety, legal and ethical issues.

Finally, the Department of Telecommunications (DoT) has formed a committee on standardisation in AI technologies to develop necessary AI standards.

3.12.4 Japan

Japan published its first AI strategy early in 2019 as discussed in [30]:

- Follow Society 5.0 principles: Cyber-Physical Society;
- Focus on "Human-Centric AI": "Dignity", "inclusivity", and "sustainability".

Several guidelines published by ministries are detailed in Fig. 3.16.

Furthermore, the Japanese government (METI/MIC) has initiated the new "AI Guideline" providing the guidelines detailed in Fig. 3.17.

Finally, Japan's approach to regulating Artificial Intelligence is summarized in Fig. 3.18.

3.12.5 United Kingdom

The overview on the **status in the UK** is given by the Department for Science, Innovation and Technology of the UK Government [31]. The objectives of the related policy debates in the UK government are specifically targeting the following objectives:

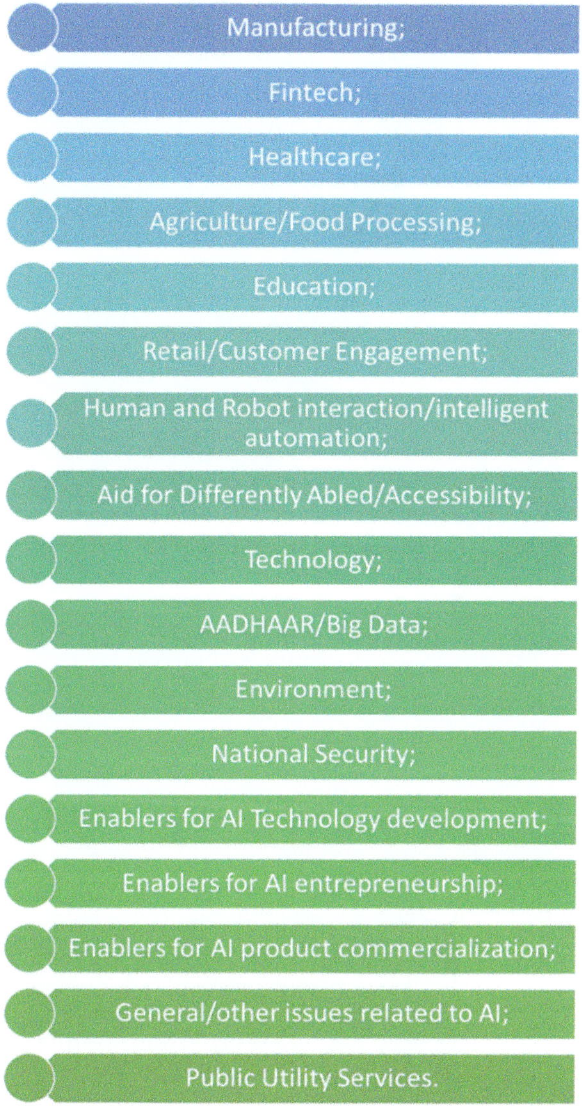

Fig. 3.15 Focus areas of "Task Force on AI for India's Economic Transformation"

- Users are protected;
- Economy can securely benefit from AI technologies;
- International alignment for security requirements.

Further, the UK approach is outlined in Fig. 3.19.

Fig. 3.16 AI Guidelines published by ministries in Japan

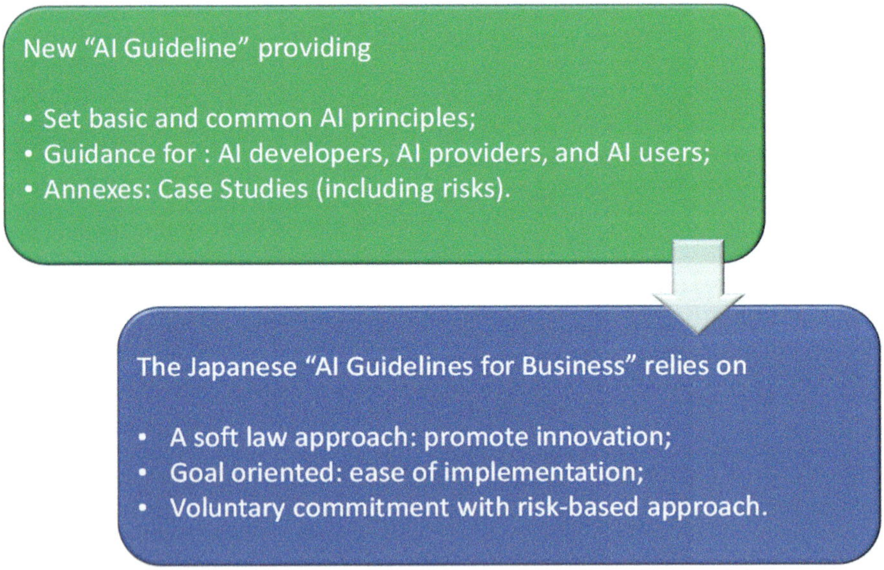

Fig. 3.17 "AI Guidelines" provided by the Japanese Government

3.12.6 Others

Canada has been developing *AI and Data Act (AIDA)* and a *Directive on Automated Decision-Making* [27].

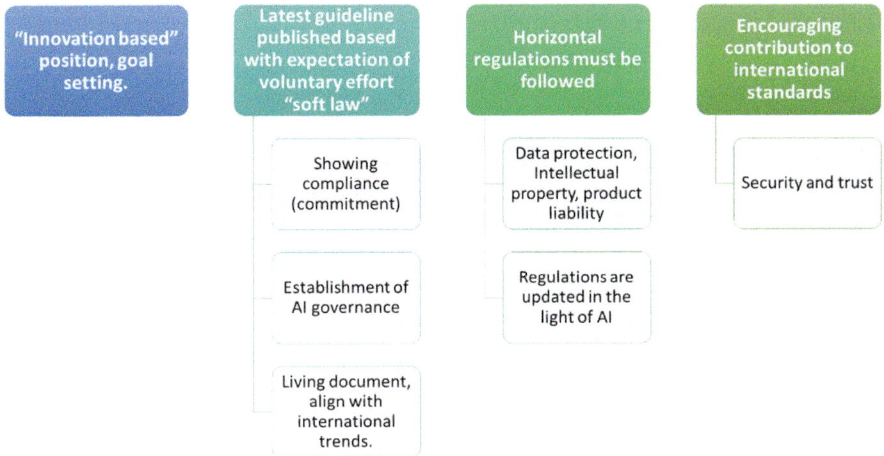

Fig. 3.18 Summary of Japan's approach to regulating Artificial Intelligence

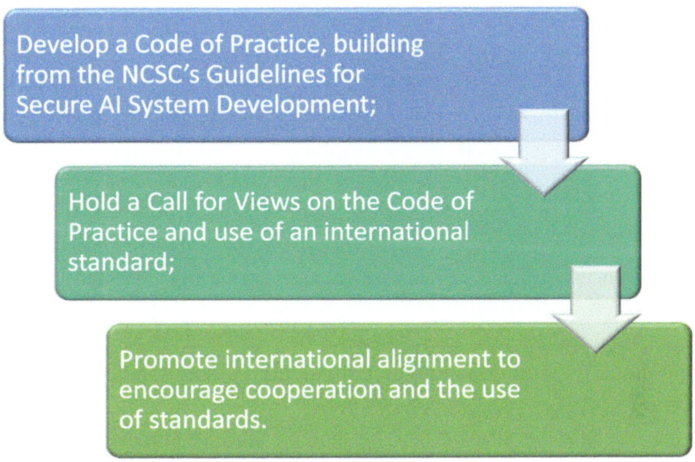

Fig. 3.19 Summary of UK approach to AI

Brazil has been developing the *AI Bill* supervised by the *Ministry of Science, Technology and Innovation, and the National Data Protection Authority* [27].

Australia AI policy lies on the foundations of the report entitled "*The State of AI Governance in Australia*" generated by the University of Technology Sydney, involving specific state authorities like *the Department of Industry, the Science and Resources, the Office of the eSafety Commissioner, the Office of the Victorian Information Commissioner* and *the Competition and Consumer Commission* [27].

X.13 Artificial Intelligence Ethics

In one of his books, Reid Blackman mentioned that "*AI products are a bit like circus tigers. You raise them like they're your own, you train them carefully, they perform beautifully in show after show after show, and then one day they bite your head off.*" Apparently, he rings a bell highlighting the importance of analyzing, designing, implementing and controlling an AI framework in which AI will be free to perform independently in order to be able to contribute to the society. This task also requires the existence of an AI ethical framework which will provide reassurances that the use of AI will be controlled [32].

According to SAP, "*AI ethics refers to the principles that govern AI's behavior in terms of human values. AI ethics helps ensure that AI is developed and used in ways that are beneficial to society. It encompasses a broad range of considerations, including fairness, transparency, accountability, privacy, security, and the potential societal impacts.*" [33].

UNESCO's 2030 Agenda for Sustainable Development mentions that "*Artificial intelligence is enabling the development of invaluable services and taking part in more and more aspects of our lives. Built from data, hardware and connectivity, AI allows machines to mimic human intelligence such as perception, problem-solving, linguistic interaction or creativity.*" [34].

Under UNESCO mandate, the AI ethical framework need is addressed by introducing a recommendation on the **Ethics of Artificial Intelligence** incorporating ethical issues regarding AI. It gives emphasis to the framework's evolutionary process, which consists of morals, values, principles, and actions that help societies navigate the effects of AI on various aspects of life. This manages to establish the context of ethical implications of AI technology, which can perform sophisticated data processing and produce artificially intelligent behavior. All phases of the AI life cycle are addressed, including monitoring, research, development, implementation, and termination. It highlights various actors, roles, and procedures including the impact to education, culture, and science under the scope of human rights, bias, and discrimination [35].

To address the AI ethics issues, SAP suggests the presence of a series of principles including *human wellbeing and dignity, human oversight, addressing bias and discrimination, transparency and explainability, upholding data privacy and protection, promoting inclusivity and diversity, society and economies, enhancing digital skills and literacy, and the health of businesses.* This would require different roles in AI ethics including *developers, researchers, policymakers, regulators, business executives, industry leaders, civil society organizations, academia, and end users* [33].

It also introduces an AI ethics notions and terms vocabulary which is essential for the reader to discuss the ethics of AI and comprehend the entities of such a technological advancement including *AI, AI ethics, AI model, AI system, agency, bias, explainability, fairness, human-in-the-loop, interpretability, large language model (LLM), machine learning (ML), normative, and transparency.* The formation of an AI steering committee is important and vital for providing a high level of accountability and oversight to the organizations.

Being responsible for the ethical issues it provides certain advantages including *composition and expertise, the purpose, scope, roles and responsibilities definition, setting objectives, creating procedures and preserves ongoing education and adaptation.* Regarding the creation of an AI ethics policy it introduces an initial approach which incorporates certain stages like *drafting the initial policy, consultation and input, integration of interdisciplinary insights, defining High Risk and red line use cases, review and revisions, finalization and approval, implementation and ongoing oversight* [33].

OECD AI Principles refer to principles for trustworthy AI. They were initially formed by OECD in 2019 and updated in 2024. The current principle serves the cause of guiding AI actors to implement trustworthy AI rules enabling policymakers to use ethical and effective AI policies. According to OECD, there are countries that use the OECD AI Principles and other related tools to form AI risk policies and frameworks. Among them, the European Union, the European Commission, the United States, and the United Nations make use of the OECD's AI principles in their AI regulatory and legislation frameworks. The basic AI principles, as referred to by the OECD organization, are distinguished into two categories, the **values-based principles** and the **recommendations for policy makers** [36].

Values Based Principles:

1. *Inclusive growth, sustainable development and well-being;*
2. *Human rights and democratic values, including fairness and privacy;*
3. *Transparency and explainability;*
4. *Robustness, security and safety;*
5. *Accountability* [36].

Recommendations for Policy Makers Principles:

1. *Investing in AI research and development;*
2. *Fostering an inclusive AI-enabling ecosystem;*
3. *Shaping an enabling interoperable governance and policy environment for AI;*
4. *Building human capacity and preparing for labour market transition;*
5. *International co-operation for trustworthy AI* [36].

Concerning the AI introduction to societies, AI seems to be gradually gaining ground in people's daily lives, affecting societies in various ways. Trivial questions refer to the vision, mission, scope, nature, cause, and the 'black boxes' of AI implications. Many organizations are alerted, aiming to take all the precautionary measures required, under the form of AI initiatives, to define ethical principles that will allow for a safe transition to an AI era. Nevertheless, it seems that the vast number of initiatives, along with their principles, tends to become overwhelmingly large in number and complex. Hence the generation of a unified framework of five principles for ethical AI can provide the clear initial map for navigation. The **five principles** refer to [37].

Beneficence: Promoting well-being, preserving dignity, and sustaining the planet, Non-maleficence: Privacy, security and 'capability caution', Autonomy: The power to decide (to decide), Justice: Promoting prosperity, preserving solidarity, avoiding unfairness and *Explicability: Enabling the other principles through intelligibility and accountability* [37].

Societies around the world have or are about to incorporate AI technology in their state services. There are a series of AI applications examples in the public sector including health and social care, education, local government, energy and utilities, transport, environment and agriculture, defense and security, criminal justice, immigration and policing, digital markets and communications, across domains, and government [38].

According to the article "Global AI Regulations Tracker: Europe, Americas & Asia–Pacific Overview" there are many states or unions of states which have developed an AI regulatory framework incorporating an ethical regulatory framework. Europe and the European Union including the United Kingdom and Switzerland, the United States, Canada, Brazil, China, Japan, India, and Australia, have initiated a series of AI regulatory authorities, services, frameworks, bills and initiatives [27].

The European Union along with the United States and the United Kingdom seem to have taken under serious consideration the potential impact of AI in their societies thus they have adapted and developed adequate to this situation precautionary measures. In an effort to focus on the European Union AI regulation and ethical framework (AI Act) the authors managed to gather the appropriate significant information which refer to specific AI Act articles regarding sensitive ethical issues.

In an effort to skyrocket research and industrial progress while reassuring safety and fundamental human and animal rights, the EU's approach to AI is focused on creating a set of rules known for their competence and trust. The EU's AI approach will allow the world to define the future while building an adequate Europe for the upcoming Digital Decade where people and businesses will be able to taste the fruits of AI in a safe and protected environment. The European AI Strategy focuses on upgrading the EU to a world-class AI hub while ensuring the human-centric and trustworthy nature of AI through concrete rules and actions [39].

Over the last decade, AI regulation has signified a central policy pillar in the EU. Policymakers have aimed to implement a 'human-centric' nature of AI to ensure that European citizens can benefit from new technological advancements which are built based on the EU's values and principles [40].

In 2020, the European Commission through its AI White Paper promised to communicate the progress of AI and promote the emerging risks with specific uses of this new technology. The initial adoption of a soft-law approach with the publication entitled "*2019* ***Ethics Guidelines for Trustworthy AI and Policy and investment recommendations***" led the European Commission change course towards a new regulation context, inviting all the interested parties for the adoption of harmonised set of rules for the implementation, communication, and use of AI technologies [41].

The **AI High-Level Expert Group (AI HLEG)** made the publication of the **Ethics guidelines for trustworthy AI** first draft on 8 April 2019. Basen on the

Guidelines document, trustworthy AI should be governed by three major components [41]:

1. *It must be lawful—respecting all applicable laws and regulations;*
2. *It must be* ***ethical****—respecting ethical principles and values;*
3. *It must be robust—both from a technical perspective while taking into account its social environment.*

According to the fundamental rights and ethical principles, the **Ethics Guidelines** consists of seven key obligations that AI systems should follow to be trustworthy [41]:

1. *Human agency and oversight;*
2. *Technical robustness and safety;*
3. *Privacy and Data governance;*
4. *Transparency;*
5. *Diversity, non-discrimination and fairness;*
6. *Societal and environmental well-being;*
7. *Accountability.*

The AI HLEG published the final version of the ***Assessment List for Trustworthy AI*** **(ALTAI)** in July 2020. ALTAI refers to a practical and useful tool that helps the reader comprehend and interpret the Ethics Guidelines into a useful set of guidelines. According to **ALTAI**: "*Achieving Trustworthy AI requires not only compliance with the law, which is but one of its three components. Laws are not always up to speed with technological developments, can at times be out of step with ethical norms or may simply not be well suited to addressing certain issues. For AI systems to be trustworthy, they should hence also be ethical, ensuring alignment with ethical norms.*" [41, 42].

The authors provide the most **AI Act ethics** related recitals and articles.

Recital 7—High Risk AI systems explains that common rules for High Risk AI systems should be established to address the protection of public interests including health, safety and fundamental rights, etc. Furthermore, ***Ethics guidelines for trustworthy AI*** should be considered as developed by the High-Level Expert Group on Artificial Intelligence (AI HLEG) [3, 43].

Recital 8—Framework for harmonised rules emphasizes the importance of and support for Small and Medium Enterprises including start-ups. There is the ambition for the EU to become a global leader in the development of secure, trustworthy and ethical AI [3, 43].

Recital 25—No application to scientific research and development addresses exemptions for research and development from the application from specific provisions of the EU AI Act [3, 43].

Recital 27—Principles for trustworthy and ethical AI refers to and reminds of the 2019 Ethics guidelines for trustworthy AI developed by the independent AI High-Level Expert Group on Artificial Intelligence (AI HLEG) [3, 43].

Recital 165—Voluntary codes of conduct states that there is the expectation for a larger uptake of ethical and trustworthy AI in the EU. Stakeholders should be encouraged to apply on a voluntary basis additional requirements related, for example, to the elements of the Union's ***Ethics Guidelines for Trustworthy AI***, environmental sustainability and others.

Article 60 AI Act—Testing of High Risk AI systems in real world conditions outside AI regulatory sandboxes, addresses the testing of High Risk AI systems and requires that it should be executed without *prejudice to any ethical review that is required by Union or national law* [3, 43].

Article 95 AI Act—Codes of conduct for voluntary application of specific requirements require the AI Office and the Member States to develop codes of conduct including Applicable elements provided for in Union ethical guidelines for trustworthy AI [3, 43].

According to Fig. 3.20, the number of scientific articles regarding ethical AI frameworks inclines. The implementation of an ethical AI framework becomes essential. This framework can protect persons, groups of people, societies and businesses in a productive way that will benefit humanity. Facing a rapid evolution of AI and recognizing the importance of ethical frameworks presence, companies, stakeholders and people organizations are actively participating in the formation of a more complete ethical framework for AI. Their objective is to identify the values that must be protected, and how they can be enforced [45].

AI Ethics will distinguish what is morally right or wrong, good or evil, aiming to guide and protect the society under the constitutional rights to justice, equal right and well fear. Within the AI Act, AI ethics has the chance to define the context and the rules under which IT developers, hardware manufacturers, legal authorities and users will operate in order to eliminate the risk of making use of unethical AI [46].

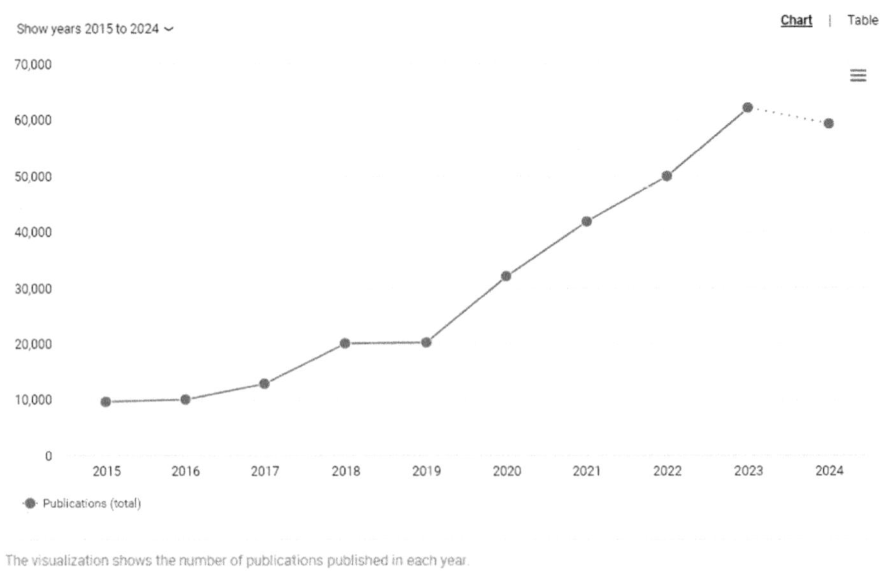

Fig. 3.20 Scientific articles regarding ethical AI frameworks [44]

According to Reuters, the big technological giants push to dilute Europe's AI Act. Thus, the EU had invited companies, academics, and other experts from the field of AI to assist in drafting the initial code of practice. The number of applications reached 1000 applications. A cluster of businesses have criticized the EU for putting tech regulation first over innovation and this will lead to difficulties in compromising when it gets to the text of the code of practice. It is also mentioned that, by the time the code is published, the companies will have a period of until August 2025 to appeal. In the race of drafting the initial code there were participating some non-profit organizations, like Access Now, the Future of Life Institute, etc. One of the biggest tech players did also apply to participate in code drafting [47].

Among others, there were more than 150 University Professors from European Universities who have asked the European institutions to incorporate an initial rights impact assessment in the AI regulation.

The academics demanded the following measures to be included in the AI Act:

1. *Clear parameters about the assessment of the impact of AI on fundamental rights;*
2. *Transparency about the results of the impact assessment through public meaningful summaries;*
3. *Participation of affected end-users, especially if in a position of vulnerability;*
4. *Involvement of independent public authorities in the impact assessment process and/or auditing mechanisms* [48].

3.13 Conclusion

The present chapter provides an overview on the recently published EU AI Act [3]. All information has been collected to the best knowledge of the authors. In this context, it is understood that the only legally binding text is the one provided in the documents formally published in the European Union Official Journal.

The EU AI Act is a major regulation which is expected to have far reaching impact on related debates in other regions. The implementation of the EU AI Act is a multi-years process and all stakeholders will need to discover its evolution carefully. It is expected that further standardisation mandates will be issued to European Standardisation Organisations leading to Harmonised European Norms—or Harmonised Standards—which can be used for presumption of conformity by manufacturers of AI systems. Also, the EU AI Act introduces the possibility for guidelines which will likely be published in the future and provide further details on the interpretation of the EU AI Act.

All stakeholders are recommended to engage in the implementation process, especially through participation in the standards making process in European Standardisation Organisations. It is a unique opportunity to shape the definition of implementation and validation requirements through Harmonised European Norms—or Harmonised Standards.

Acknowledgements The authors express specific thanks to Intel colleague Mario Romao who has substantially supported the writing of this chapter through providing valuable advice and careful proof-reading.

Disclaimer Note that the authors have assembled the information above to the best of their knowledge as available at the time when the present book was written. Any interpretation is the personal opinion of the authors. Also, the regulation framework can change and evolve any time. The only legally binding text is the one provided in the documents formally published in the European Union Official Journal.

References

1. Proposal for a REGULATION OF THE EUROPEAN PARLIAMENT AND OF THE COUNCIL LAYING DOWN HARMONISED RULES ON ARTIFICIAL INTELLIGENCE (ARTIFICIAL INTELLIGENCE ACT) AND AMENDING CERTAIN UNION LEGISLATIVE ACTS, 21st April 2021, European Commission
2. Markus Dominik Mueck, Amit Elazari Bar On, Stephane Du Boispean, Upcoming European Regulations on Artificial Intelligence and Cybersecurity, IEEE Communications Magazine, Volume: 61, Issue: 7, July 2023
3. REGULATION (EU) 2024/1689 OF THE EUROPEAN PARLIAMENT AND OF THE COUNCIL of 13 June 2024 laying down harmonised rules on artificial intelligence and amending Regulations (EC) No 300/2008, (EU) No 167/2013, (EU) No 168/2013, (EU) 2018/858, (EU) 2018/1139 and (EU) 2019/2144 and Directives 2014/90/EU, (EU) 2016/797 and (EU) 2020/1828 (Artificial Intelligence Act)
4. 1st European AI Office webinar, European AI Office, 30th May 2024, available at https://digital-strategy.ec.europa.eu/en/events/1st-european-ai-office-webinar-risk-management-logic-ai-act-and-related-standards
5. The Blue Guide, European Commission, 2022, available at https://eur-lex.europa.eu/legal-content/EN/TXT/?uri=OJ:C:2022:247:TOC
6. COMMISSION IMPLEMENTING DECISION on a standardisation request to the European Committee for Standardisation and the European Committee for Electrotechnical Standardisation in support of Union policy on artificial intelligence, 22nd May 2023, available at https://ec.europa.eu/transparency/documents-register/detail?ref=C(2023)3215&lang=en
7. The Vienna Agreement, CEN/CENELEC, available at https://www.cencenelec.eu/about-cen/cen-and-iso-cooperation/
8. The Frankfurt Agreement, CEN/CENELEC, available at https://www.cencenelec.eu/about-cenelec/cenelec-and-iec-cooperation/
9. ISO/IEC 42001:2023, Information technology — Artificial intelligence — Management system, available at https://www.iso.org/standard/81230.html
10. Standardisation work in support of AI Act Antoine-Alexandre André - DG CNECT A.2, ETSI AI Conference, February 2024, available at https://docbox.etsi.org/Workshop/2024/02_ETSIAICONFERENCE/S01_OPENING
11. European AI Office, available at https://digital-strategy.ec.europa.eu/en/policies/ai-office
12. Definition of Risk, Wikipedia, available at https://en.wikipedia.org/wiki/Risk
13. ISO 9001:2015: Quality management systems — Requirements, Edition 5, 2015, available at https://www.iso.org/standard/62085.html#:~:text=ISO%209001%20is%20a%20globally,demonstrate%20their%20commitment%20to%20quality.
14. Directive (EU) 2022/2557 of the European Parliament and of the Council of 14 December 2022 on the resilience of critical entities and repealing Council Directive 2008/114/EC (Text with EEA relevance), available at https://eur-lex.europa.eu/eli/dir/2022/2557/oj

15. INTERPLAY BETWEEN THE RADIO EQUIPMENT DIRECTIVE AND THE ARTIFICIAL INTELLIGENCE ACT, EUROPEAN COMMISSION DIRECTORATE-GENERAL FOR INTERNAL MARKET, INDUSTRY, ENTREPRENEURSHIP AND SMES, submission to 19th EC Expert Group Radio Equipment, EG RE (19)21, 2024, available at https://circabc.europa.eu/ui/group/43315f45-aaa7-44dc-9405-a86f639003fe/library/bdfad6eb-8934-4b55-865d-ecda2695f2d5?p=1&n=10&sort=modified_DESC
16. AI Watch, AI standardisation landscape state of play and link to the EC proposal for an AI regulatory framework, Joint Research Center of the EC, 2021, available at https://op.europa.eu/en/publication-detail/-/publication/36c46b8e-e518-11eb-a1a5-01aa75ed71a1/language-en
17. Status of JTC 21 Activities in Response of the Standardization Request on AI, CEN and CENELEC presentation at ETSI AI Conference, February 2024, available at https://docbox.etsi.org/Workshop/2024/02_ETSIAICONFERENCE/S03_STANDARDIZ_AIACT_LEGALFMK/JTC21ACTIVITIES_STD_REQUESTAI_CENCENELECJTC21.pdf
18. Current and Future Activities of ETSI TC SAI (Securing Artificial Intelligence), Scott CADZOW, Chair ETSI Technical Committee Securing Artificial Intelligence (TC SAI), ETSI AI Conference, February 2024, available at https://docbox.etsi.org/Workshop/2024/02_ETSIAICONFERENCE/S04_CYBER_SECURITY/ETSI_TC_SAI_CADZOW.pdf
19. Regulation (EU) 2019/881 of the European Parliament and of the Council of 17 April 2019 on ENISA (the European Union Agency for Cybersecurity) and on information and communications technology cybersecurity certification and repealing Regulation (EU) No 526/2013 (Cybersecurity Act) (Text with EEA relevance), available at https://eur-lex.europa.eu/legal-content/EN/TXT/?uri=celex%3A32019R0881
20. Proposal for a DIRECTIVE OF THE EUROPEAN PARLIAMENT AND OF THE COUNCIL on adapting non-contractual civil liability rules to artificial intelligence (AI Liability Directive), available at https://eur-lex.europa.eu/legal-content/EN/TXT/?uri=CELEX%3A52022PC0496
21. Liability Rules for Artificial Intelligence, European Commission, available at https://commission.europa.eu/business-economy-euro/ng-business-eu/contract-rules/digital-contracts/liability-rules-artificial-intelligence_en
22. Proposal for a directive of the European Parliament and of the Council on Liability for defective products, (COM(2022)0495 – C9-0322/2022 – 2022/0302(COD)) available at https://view.officeapps.live.com/op/view.aspx?src=https%3A%2F%2Fwww.europarl.europa.eu%2FRegData%2Fcommissions%2Fimco%2Finag%2F2024%2F01-24%2FCJ24_AG(2024)758731_EN.docx&wdOrigin=BROWSELINK
23. Tech Law Blog, Beyond the AI Act: The AI Liability Directive & the Product Liability Directive, Julia Launders, March 2024, available at https://www.techlaw.ie/2024/03/articles/artificial-intelligence/beyond-the-ai-act-how-the-ai-liability-directive-and-the-product-liability-directive-will-also-shape-the-regulation-of-ai-in-the-eu/
24. EU-US Trade and Technology Council, available at https://commission.europa.eu/strategy-and-policy/priorities-2019-2024/stronger-europe-world/eu-us-trade-and-technology-council_en
25. Executive Order on the Safe, Secure, and Trustworthy Development and Use of Artificial Intelligence, OCTOBER 30, 2023; US White House; available at https://www.whitehouse.gov/briefing-room/presidential-actions/2023/10/30/executive-order-on-the-safe-secure-and-trustworthy-development-and-use-of-artificial-intelligence/
26. Department of Commerce Announces New Actions to Implement President Biden's Executive Order on AI, available at https://www.nist.gov/news-events/news/2024/04/department-commerce-announces-new-actions-implement-president-bidens
27. Legal Nodes, Global AI Regulations Tracker: Europe, Americas & Asia-Pacific Overview, Kostiantyn Ponomarov, June 24 2024, available at https://legalnodes.com/article/global-ai-regulations-tracker
28. China AI Policies and Standardizations, Dr. Betty Xu, ETSI AI Conference, February 2024, available at https://docbox.etsi.org/Workshop/2024/02_ETSIAICONFERENCE/S02_REGULATION_GLOBAL/CHINA_SESEC_BETTY_XU.pdf

29. AI Regulation and Standardization in India, Dinesh Chand Sharma, ETSI AI Conference, February 2024, available at https://docbox.etsi.org/Workshop/2024/02_ETSIAICONFERENCE/S02_REGULATION_GLOBAL/INDIA_SESEI_DINESH_SHARMA.pdf
30. Overview of the Japanese position on AI, Francois Ortolan, ETSI AI Conference, February 2024, available at https://docbox.etsi.org/Workshop/2024/02_ETSIAICONFERENCE/S02_REGULATION_GLOBAL/JAPAN_OVERVIEW_AI_NEC__ORTOLAN.pdf
31. The UK Government's Approach to the Cyber Security of AI, James Deacon, ETSI AI Conference, February 2024, available at https://docbox.etsi.org/Workshop/2024/02_ETSIAICONFERENCE/S02_REGULATION_GLOBAL/UK_DIST_DEACON.pdf
32. Blackman, Reid, and Jim Euchner. Ethics in the Age of AI: A Conversation with Reid Blackman. Research-Technology Management, Volume: 67, Issue: 1, Pages: 15–21, https://doi.org/10.1080/08956308.2024.2280481
33. What is AI ethics?, August 2024, available at https://www.sap.com/resources/what-is-ai-ethics
34. Artificial Intelligence, August 2024, available at https://www.unesco.org/en/artificial-intelligence
35. UNESCO, Recommendation on the Ethics of Artificial Intelligence, November 2021, available at https://unesdoc.unesco.org/ark:/48223/pf0000381137
36. OECD.AI, OECD AI Principles overview, May 2024, available at https://oecd.ai/en/ai-principles
37. Luciano Floridi, Josh Cowls, A Unified Framework of Five Principles for AI in Society. In: Floridi, L. (eds) Ethics, Governance, and Policies in Artificial Intelligence. Philosophical Studies Series, Volume 144. Springer, Cham, 2021, https://doi.org/10.1007/978-3-030-81907-1_2
38. The Alan Turing Institute, AI Ethics and Governance in Practice, AI Ethics and Governance in Practice an Introduction, 2024, available at https://aiethics.turing.ac.uk/modules/introduction/?modulepage=2
39. European Commission, Shaping Europe's digital future, European approach to artificial intelligence, August 1 2024, available at https://digital-strategy.ec.europa.eu/en/policies/european-approach-artificial-intelligence
40. European Parliament, BRIEFING EU, Legislation in Progress, Artificial intelligence act, 2024, available at https://www.europarl.europa.eu/RegData/etudes/BRIE/2021/698792/EPRS_BRI(2021)698792_EN.pdf
41. European Commission, Shaping Europe's digital future, Ethics guidelines for trustworthy AI, January 30 2024, available at https://digital-strategy.ec.europa.eu/en/library/ethics-guidelines-trustworthy-ai
42. European Commission, Shaping Europe's digital future, Ethics guidelines for trustworthy AI, July 17 2024, available at https://ec.europa.eu/newsroom/dae/document.cfm?doc_id=68342, https://doi.org/10.2759/002360
43. Intersoft Consulting, AI Act, October 2024, available at https://ai-act-law.eu/
44. Digital Science's Dimensions, AI ethics frameworks, September 22 2024, available at https://app.dimensions.ai/analytics/publication/overview/timeline?search_mode=content&search_text=ai%20ethics%20framework&search_type=kws&search_field=full_search
45. Dimnesions.ai, AI ethics frameworks, available at https://app.dimensions.ai/analytics/publication/overview/timeline?search_mode=content&search_text=ai%20ethics%20framework&search_type=kws&search_field=full_search
46. Medium, The Need for an Ethical Approach to AI, December 19 2023, available at https://heka-ai.medium.com/the-need-for-an-ethical-approach-to-ai-8d8993a74586
47. Reuter, Tech giants push to dilute Europe's AI Act, September 20 2024, available at https://www.reuters.com/technology/artificial-intelligence/tech-giants-push-dilute-europes-ai-act-2024-09-20/?utm_source=substack&utm_medium=email
48. Brussles Privacy Hub, September 12 2023 , available at https://brusselsprivacyhub.com/2023/09/12/brussels-privacy-hub-and-other-academic-institutions-ask-to-approve-a-fundamental-rights-impact-assessment-in-the-eu-artificial-intelligence-act/?utm_source=substack&utm_medium=email

Chapter 4
Introduction to the European Cyber Resilience Act

Markus Mueck, Taylor Roberts, Stéphane Du Boispéan, and Christophe Gaie

Abstract In September 2022, a proposal was published *for a regulation of the European Parliament and Council on horizontal cybersecurity requirements for products with digital elements, amending Regulation (EU) 2019/1020* on Market Surveillance and Compliance of Products. While the present book was written, the legislative procedure was still ongoing and the latest available Draft was published in October 2024 as the version adopted by the European Council. The proposed *Cyber Resilience Act (CRA)* represents a major change of regulating access of *products with digital elements*—which relates to (almost) all of today's digital components and products including their remote data processing solutions, consumer products, etc.—to the European Union Single Market. New security requirements, specific product life-cycle obligations, and new vulnerability/incident reporting requirements are introduced for placement of such products on the European Union Single Market. Beyond Europe, the CRA is expected to influence cybersecurity related debates globally. The CRA will also likely have substantial impact on the implementation of existing cybersecurity regulation in the EU, including the Radio Equipment Directive and the Network and Information Security 2 (NIS2) Directive. We propose an overview of the draft regulation and discuss possible implementation solutions for compliance with the new essential requirements with the objective to main (continued) access to the European Union Single Market.

M. Mueck (✉)
Intel Deutschland GmbH, Munich, Germany
e-mail: Markus.Dominik.Mueck@intel.com

T. Roberts
Intel UK, Swindon, UK
e-mail: Taylor.Roberts@intel.com

S. Du Boispéan
Intel Corporation SA/NV, Kontich, Belgium
e-mail: Stephane.du.Boispean@intel.com

C. Gaie
Head of the Engineering and Digital Innovation Division, Prime Minister Services, Paris, France
e-mail: christophe.gaie@gmail.com

M. Mueck and C. Gaie (eds.), *European Digital Regulations*, Intelligent Systems Reference Library 265, https://doi.org/10.1007/978-3-031-80809-8_4

Keywords Regulation · Digital · Artificial intelligence · Machine learning · European Union single market

4.1 Introduction

On 15th September 2022, the original Draft of the Cyber Resilience Act (CRA) [1] was published followed by an updated version published in December 2023 [2]; a related draft standardization request (SR) [3] followed on 17th April 2024 and was subsequently updated on 20th September 2024 and made available as a new draft version [4]. While the present book was written, the legislative procedure was still ongoing and the latest available Draft of the Cyber Resilience Act (CRA) was published on October 2024 as the version adopted by the European Council [5]. A corresponding press release was published by the European Council [6]. The official publication in the European Union Official Journal is expected in the weeks afterwards. Some scientific literature is already published to explain its impacts on economic operators and articulation with other regulations [7, 8].

The Cyber Resilience Act will have major implications on the access of products to the European Union Single Market. Indeed, Cybersecurity related requirements will become mandatory for so-called *products with digital elements* which is defined as follows by the draft CRA [5]:

'product with digital elements' means a software or hardware product and its remote data processing solutions, including software or hardware components being placed on the market separately

An illustration of the objectives of the CRA is proposed in Fig. 4.1.

The regulation thus not only relates to finished products to be placed onto the European Union Single Market, but also to *software or hardware components.* For the manufacturers of such components, the corresponding requirements are very novel and important to be understood; in case of non-compliance, access to the European Union Single Market may be lost. While regulations such as the Radio Equipment Directive [9] are limited to the placement of products onto the European

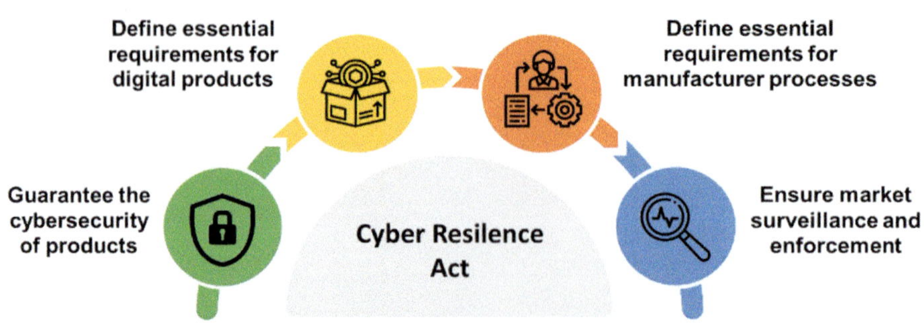

Fig. 4.1 Main objectives of the cyber resilience act

Union Single Market (i.e., the essential requirements need to be met at the specific point in time when the product is placed onto the market as defined by the *Blue Guide* [10]; afterwards, once the product is placed onto the market, other mechanisms apply such as market surveillance, etc.), the draft CRA (i) addresses the placement onto the European Union Single Market and furthermore (ii) relates to the management of the product over its lifetime (no less than 5 years unless the lifetime is shorter). The requirement to maintain compliance over the lifetime of the product or component (no less than 5 years unless the lifetime is shorter) is thus very new and has deep consequences for manufacturers.

It is also noteworthy that recently a Commission Delegated Regulation [11] was issued for the activation of Cybersecurity and Privacy related articles of the Radio Equipment Directive [9]. Obviously, there is an overlap in scope with the draft CRA. It is therefore anticipated that the RED Articles on Cybersecurity will be moved under the CRA, i.e. … *the Commission repeals or amends Delegated Regulation (EU) 2022/30* [11]… Corresponding provisions are included in the Draft CRA text [5].

4.2 Implementation Time-Line

At the time when the present chapter was written, the finalization of the Cyber Resilience Act (CRA) was still ongoing and the exact time of the Entry Into Force (EIF) was not yet known, but was expected for late 2024. Once the CRA is published in the European Union Official Journal (EUOJ), the milestones for implementation will follow this timeline [12, 13], as illustrated in Fig. 4.2.

Note that the timing of the "Application of Art. 14" was updated in Fig. 4.1 according to the latest available draft of the CRA [5]. The original Draft CRA was indicating a "Entry Into Force (EIF) + 12 months" time line while this was increased

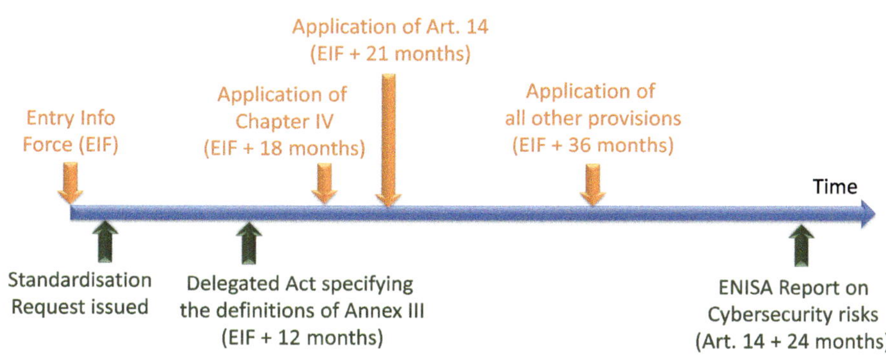

Fig. 4.2 Expected implementation time-line of the cyber resilience act (CRA) (timeline taken from [13] and modified to reflect the timing definitions of the latest draft CRA version [5])

to "EIF + 21 months" in its latest version [5]. Similary, the date of application of all other provisions moved from "EIF + 24 months" [1] to "EIF + 36 months" [5].

There are a number of security requirements that are referenced in the body of the Draft CRA, with further clarity on their scope and context to be included in the Implementing Act and subsequent standards. Article 14 mentioned in the illustration above relates to Reporting obligations of manufacturers. To summarize, as illustrated in Fig. 4.3, these include.

Furthermore, the Cyber Resilience Act defines the concepts illustrated in Fig. 4.4.

The authors expect that further discussion will be required on which type of incidents or vulnerabilities will finally need to be reported. In practice, numerous vulnerabilities may occur which have no practical impact and thus extensive reporting may not lead to a significant added value for such cases; on the other hand, it may create considerable cost for manufacturer.

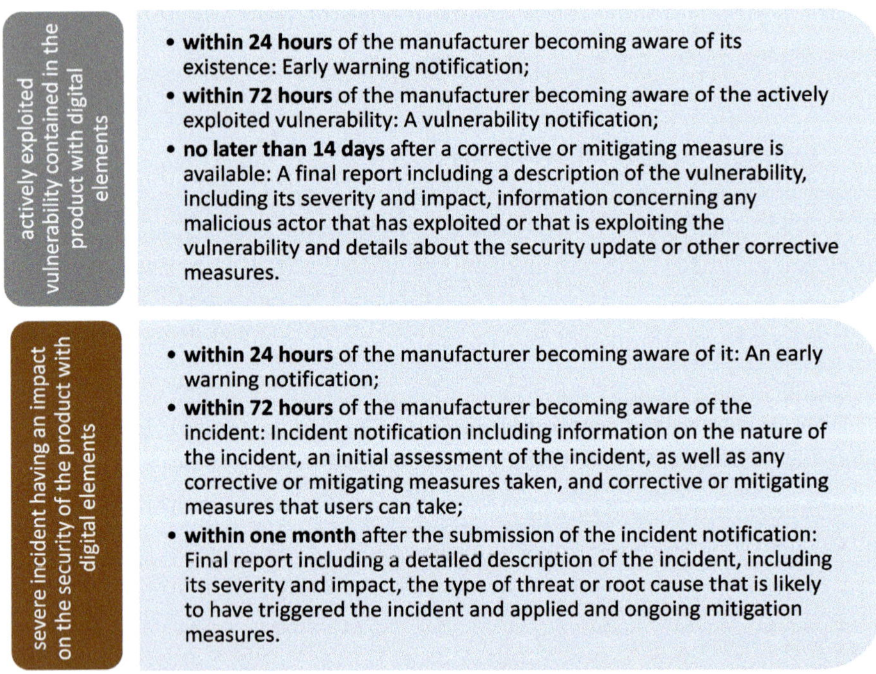

Fig. 4.3 Reporting requirements under the cyber resilience act

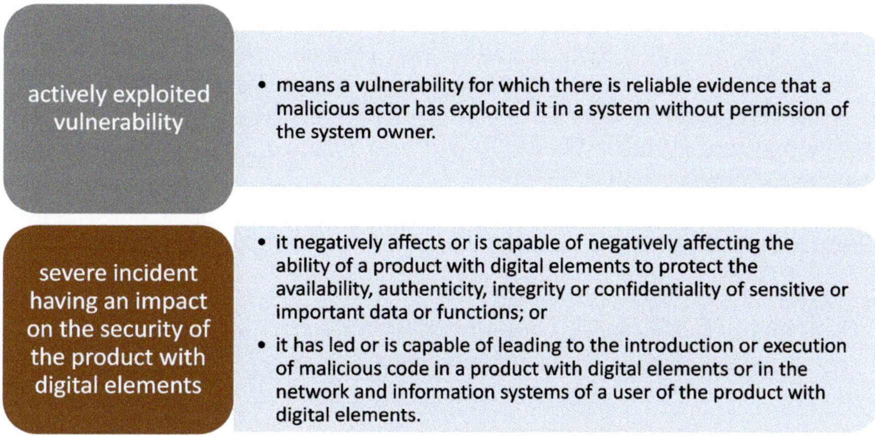

Fig. 4.4 Definitions of actively exploited vulnerability and severe incidents

4.3 Overview on Essential Requirements Introduced by the Cyber Resilience Act

Annex I of the Draft Cyber Resilience Act (CRA) introduces the essential requirements to be met by manufacturers to achieve access to the European Union Single Market for products with digital elements. A summary of the essential requirements is given in Fig. 4.5.

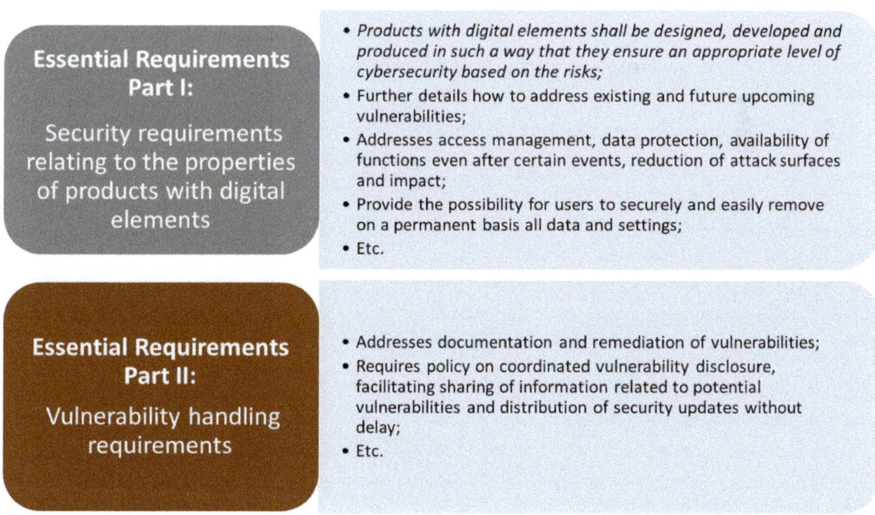

Fig. 4.5 Overview on essential requirements introduced by the Draft CRA for products with digital elements

As is typically the case for Regulations under the New Legislative Framework (NLF), a Standardisation Request will be issued providing further guidance on the implementation of the Essential Requirements summarized above. At the time when the present chapter was written, a Draft Standardisation Request [3] and a further revision [4] were available for the Draft Cyber Resilience Act (CRA) and were requiring the development of the following horizontal deliverables (i.e., applicable to all products with digital elements)—the corresponding titles give indeed a good indication on how the upper Essential Requirements should be understood:

Draft Cyber Resilience Act Standardisation Request [4] Annex I

Horizontal standards for security requirements relating to the properties of products with digital elements: European standard(s) on

- *designing, developing and producing products with digital elements in such a way that they ensure an appropriate level of cybersecurity based on the risks;*
- *delivering products with digital elements without any known exploitable vulnerabilities;*
- *delivering products with digital elements with a secure by default configuration;*
- *ensuring that vulnerabilities in products with digital elements can be addressed through security updates;*
- *ensuring protection of products with digital elements from unauthorised access and reporting on possible unauthorised access;*
- *protecting the confidentiality of data stored, transmitted or otherwise processed by a product with digital elements;*
- *protecting the integrity of data, commands, programs by a product with digital elements, and its configuration against any manipulation or modification not authorised by the user, as well as reporting on corruptions;*
- *processing only personal or other data that are adequate, relevant and limited to what is necessary in relation to the intended purpose of the product with digital elements ('minimisation of data');*
- *protecting the availability of essential and basic functions of the product with digital elements;*
- *minimising the negative impact of a product with digital elements or its connected devices on the availability of services provided by other devices or networks;*
- *designing, developing and producing products with digital elements with limitted attack surfaces;*
- *designing, developing and producing products with digital elements that reduce the impact of an incident using appropriate exploitation mitigation mechanisms and techniques;*

- *providing security related information by recording and/or monitoring relevant internal activity of products with digital elements with an optout mechanism for the user*;
- *securely and easily removing or transferring all data and settings of a product with digital elements*

Furthermore, the Draft Standardisation Request [4] calls for standards for vulnerability-handling requirements:

Draft Cyber Resilience Act Standardisation Request [4] Annex I

Horizontal standards for vulnerability handling requirements: European standard(s) on Vulnerability handling for products with digital elements

Finally, product-specific standards for security requirements relating to the properties of products with digital elements. The latter category includes a series of vertical standards addressing the specific needs of the following industries or applications:

Draft Cyber Resilience Act Standardisation Request [4] Annex I

Vertical standards for security requirements relating to the properties of products with digital elements: European standard(s) on

- *Essential cybersecurity requirements for identity management systems and privileged access management software and hardware, including authentication and access control readers, including biometric readers*;
- *Standalone and embedded browsers*;
- *Password managers*;
- *Software that searches for, removes, or quarantines malicious software*;
- *Products with digital elements with the function of virtual private network (VPN)*;
- *Network management systems*;
- *Security information and event management (SIEM) systems*;
- *Boot managers*;
- *Public key infrastructure and digital certificate issuance software*;
- *Physical and virtual network interfaces*;
- *Operating systems*;
- *Routers, modems intended for the connection to the internet, and switches*;
- *Microprocessors with security-related functionalities*;
- *Microcontrollers with security-related functionalities*;
- *Application specific integrated circuits (ASIC) and field-programmable gate arrays (FPGA) with security-related functionalities*;

- *Smart home general purpose virtual assistants*;
- *Smart home products with security functionalities, including smart door locks, security cameras, baby monitoring systems and alarm systems*;
- *Internet connected toys covered by Directive 2009/48/EC that have social interactive features (e.g. speaking or filming) or that have location tracking*;
- *Personal wearable products to be worn or placed on a human body that have a health monitoring (such as tracking) purpose and to which Regulation (EU) 2017/745 or Regulation (EU) 2017/746 do not apply or personal wearable products that are intended for the use by and for children*;
- *Hypervisors and container runtime systems that support virtualised execution of operating systems and similar environments*;
- *Firewalls, intrusion detection and/or prevention systems, including specifically those intended for industrial use*;
- *Tamperresistant microprocessors*;
- *Tamperresistant microcontrollers*;
- *Hardware Devices with Security Boxes*;
- *Smart meter gateways within smart metering systems as defined in Article 2 (23) of Directive (EU) 2019/944 and other devices for advanced security purposes, including for secure cryptoprocessing*;
- *Smartcards or similar devices, including secure elements*

It will obviously be a challenge for stakeholders to develop the substantial set of Harmonised European Norms outlined above. It is therefore expected to rely upon available work to the greatest extent possible, in particular work available in ETSI, CEN/CLC and ISO/IEC.

4.4 Conformity Assessment

As detailed in the European Commission Blue Guide [10], a *product has to undergo a full conformity assessment before it is made available on the market*. In case that the product is modified after placement on the market, the *related technical documentation has to be updated in as much as the modification has an impact on the requirements of the applicable legislation* [10]. The basic principle is illustrated in Fig. 4.6.

With the Cyber Resilience Act (CRA), new conformity assessment requirements are introduced beyond the basic principles illustrated above. Indeed, compliance needs to be guaranteed over the lifetime of the product (no less than 5 years unless the lifetime of the product is shorter), especially in case that new cybersecurity vulnerabilities are being identified. Corresponding mitigation solutions, typically through software upgrades, need to be made available within well-defined time frames. The basic principle is illustrated in Fig. 4.7.

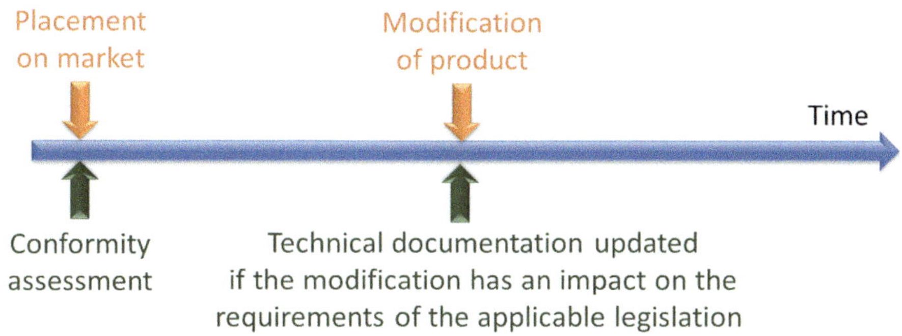

Fig. 4.6 Conformity assessment at time of placement on European union single market

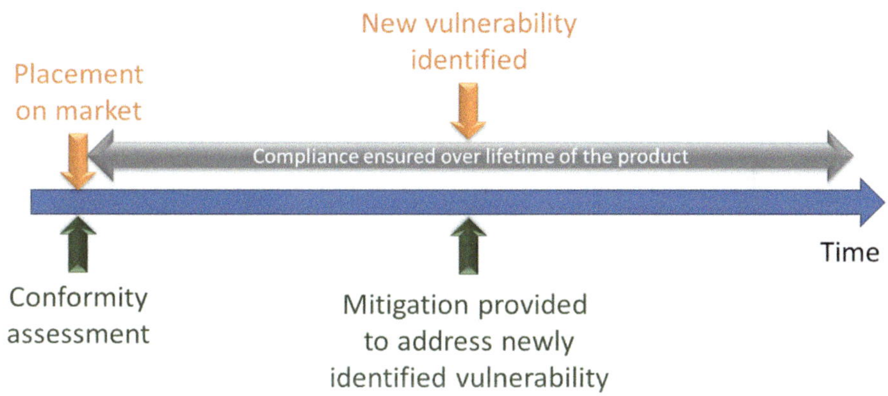

Fig. 4.7 Continued compliance over the lifetime of a product

Annex IV of the Draft CRA details conformity assessment procedures and introduces a set of "modules" which are applied for product conformity assessment depending on its classification as.

- *Important products with digital elements, Class I;*
- *Important products with digital elements, Class II;*
- *Critical products with digital elements or*
- *Other (not part of the upper classes).*

Depending on the applicable classification of a product, one or multiple of the following Modules are applicable as indicated in Fig. 4.8.

It should be noted that the last item in Fig. 4.7 creates a strong linkage between the CRA and the Cybersecurity Act, i.e. Regulation (EU) 2019/881 [14]. Still, European Cybersecurity Certification Schemes specifically tailored to CRA will need to be created in the future.

The various product classes require the application of the upper Modules and are detailed in Fig. 4.9.

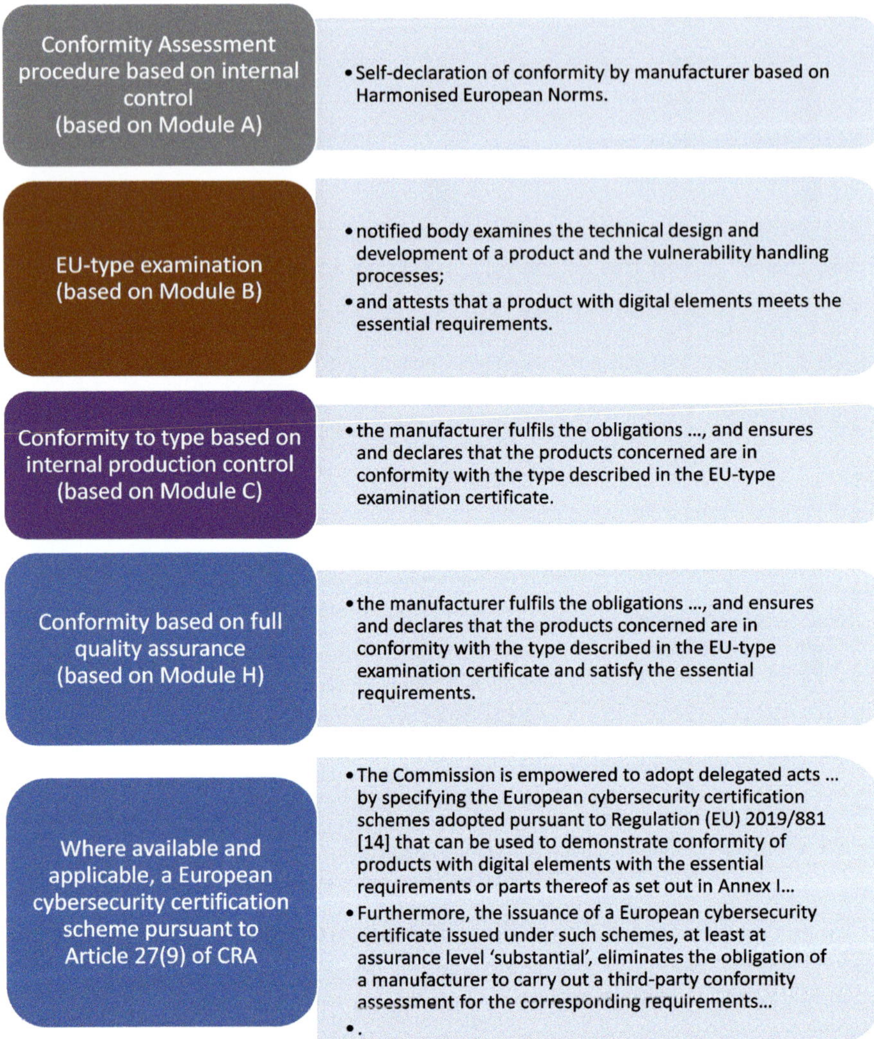

Fig. 4.8 Conformity assessment modules defined by the Draft CRA

The instructions for Critical Products with digital elements refer to Draft CRA *Article 8: Critical products with digital elements.* Article 8(1) outlines the applicable approach for the European Commission to adopt a European cybersecurity certification scheme to demonstrate conformity with the essential requirements.

Following the upper instructions for conformity assessment, it is evident that any classification as important Class II or critical product will significantly increase the complexity of obtaining product conformity assessment, since self-declaration of conformity based on Harmonised Standards is no long possible.

Product category	Conformity assessment
Neither important nor critical product with digital elements	• the internal control procedure (based on module A); or • the EU-type examination procedure (based on module B) followed by conformity to EU-type based on internal production control (based on module C); or • conformity assessment based on full quality assurance (based on module H); or • where available and applicable, a European cybersecurity certification scheme as specified in Article 27(9).
Important product with digital elements class I	• Harmonised standards, common specifications or European cybersecurity certification schemes ***at assurance level at least substantial*** can be applied by the manufacturer. • Otherwise, if the manufacturer has not applied or has applied only in part harmonised standards, common specifications or European cybersecurity certification schemes ***at assurance level at least substantial*** as referred to in Article 27, or where such harmonised standards, common specifications or European cybersecurity certification schemes do not exist, then the following applies: • EU-type examination procedure (based on module B), followed by conformity to EU-type based on internal production control (based on module C); or • conformity assessment based on full quality assurance (based on module H).
Important product with digital elements class II	• EU-type examination procedure (based on module B), followed by conformity to EU-type based on internal production control (based on module C); or • conformity assessment based on full quality assurance (based on module H); or • where available and applicable, a European cybersecurity certification scheme as specified in Article 27(9) at assurance level at least substantial pursuant to Regulation (EU) 2019/881.
Critical product with digital elements	• a European cybersecurity certification scheme in accordance with Article 8(1) of CRA; or • where the conditions in Article 8(1) of CRA are not met, any of the procedures referred to for important products with digital elements class II.

Fig. 4.9 Conformity assessment for different product categories

Important Products with digital elements are defined in Annex III for the Draft CRA [5] for Class I and Class II as follows:

Annex III

Important Products With Digital Elements

Class I

1. Identity management systems and privileged access management software and hardware, including authentication and access control readers, including biometric readers;

2. Standalone and embedded browsers;
3. Password managers;
4. Software that searches for, removes, or quarantines malicious software;
5. Products with digital elements with the function of virtual private network (VPN);
6. Network management systems;
7. Security information and event management (SIEM) systems;
8. Boot managers;
9. Public key infrastructure and digital certificate issuance software;
10. Physical and virtual network interfaces;
11. Operating systems;
12. Routers, modems intended for the connection to the internet, and switches;
13. Microprocessors with security-related functionalities;
14. Microcontrollers with security-related functionalities;
15. Application specific integrated circuits (ASIC) and field-programmable gate arrays (FPGA) with security-related functionalities;
16. Smart home general purpose virtual assistants;
17. Smart home products with security functionalities, including smart door locks, security cameras, baby monitoring systems and alarm systems;
18. Internet connected toys covered by Directive 2009/48/EC of the European Parliament and of the Council that have social interactive features (e.g. speaking or filming) or that have location tracking features;
19. Personal wearable products to be worn or placed on a human body that have a health monitoring (such as tracking) purpose and to which Regulation (EU) 2017/745 or Regulation (EU) 2017/746 do not apply or personal wearable products that are intended for the use by and for children

Class II

1. Hypervisors and container runtime systems that support virtualised execution of operating systems and similar environments;
4. Firewalls, intrusion detection and prevention systems;
5. Tamper-resistant microprocessors;
5a. Tamper-resistant microcontrollers;

Critical Products with digital elements are defined in Annex IV for the Draft CRA [2] as follows:

Annex IV

Critical Products With Digital Elements

1. Hardware Devices with Security Boxes;

2. Smart meter gateways within smart metering systems as defined in Article 2 (23) of Directive (EU) 2019/944 of the European Parliament and of the Council and other devices for advanced security purposes, including for secure cryptoprocessing;
3. Smartcards or similar devices, including secure elements;

4.5 Standards in Support of the Cyber Resilience Act

As a regulation under the New Legislative Framework (NLF), the Draft Cyber Resilience Act (CRA) allows for certain products with digital elements of limited criticality the usage of Harmonised European Norms—or Harmonised Standards—for obtaining presumption of conformity with the Cyber Resilience Act itself. Products with digital elements of higher criticality require the involvement of a Notified Body or similar further outlined above.

At the time when the present chapter was written, a corresponding Standardisation Request was available as a Draft [3, 4] but it was not yet formally issued to European Standardisation Organisations (ESOs). ESOs currently engage in preparatory activities. It is obvious that a considerable effort has to be made by all concerned stakeholders to deliver all required standards—a total of 41 are mentioned in the available Draft Standardisation Request [4].

For this purpose, it is essential to avoid reinventing the wheel and to exploit available standards to the extent possible. Indeed, the European Standardisation Organisations CEN/CENELEC have the Frankfurt [15] and Vienna [16] agreements with ISO/IEC in place which can be used to adopt corresponding standards as Harmonised Standards or other appropriate deliverable types.

Furthermore, the Joint Research Center of the European Commission (JRC) and the European Union Agency for Cybersecurity (ENISA) have conducted a *Cyber Resilience Act Requirements Standards Mapping* [17] to assess the available of standards for potential usage for the implementation of the Cyber Resilience Act. Key standards identified by this study are summarized below:

Standards identified by JRC and ENISA to support security requirements of the Cyber Resilience Act:

ISO/IEC	IEC	ITU-T	ETSI TS
EN 27002:2022	EN 62443-3-2:2020	X.1214 (03/2018)	EN 303 645 V2.1.1 (2020-06)
EN 27005:2022	EN 62443-4-1:2018	X.1253 (09/2011)	103 485 V1.1.1 (2020-08)
EN 18045:2022	EN 62443-4-2:2019	X.812 (11/1995)	
18031:2011	62443-2-1:2010	X.805 (10/2003)	
9798, Parts 1 to 6		X.814 (11/1995)	
24760, Parts 1 to 3		X.815 (11/1995)	

(continued)

(continued)

ISO/IEC	IEC	ITU-T	ETSI TS
29146:2016		Y.4810 (11/2021)	
18033, Parts 1 to 7			
9796, Parts 2 and 3			
9797, Parts 1 to 3			
14888, Parts 1 to 3			
27701:2019			
29100:2011			
22237-1:2021			
TS 19249:2017			
15408-2:2022			
27001:2022			
27034–1:2011			
EN 15408-3:2022			
13888-1:2020			
30111:2019			
Standards identified by JRC and ENISA to support vulnerability handling requirements of the Cyber Resilience Act:			
27036, Parts 1 to 3	62443-4-1:2018		303 645 V2.1.1 (2020-06)
27001:2022			
27002:2022			
EN 30111:2020			
EN 29147:2020			
TS 27034-5-1:2018			
27005:2022			

Concerning ISO/IEC standards, it should be noted that it is up to the European Standardisation Organisations CEN/CENELEC to assess the validity of the references listed above and to decide which of those should be adopted through the Frankfurt [15] and Vienna agreeements [16].

4.6 Interdependencies with Other European Regulations

4.6.1 Cybersecurity Act (CSA)

There is a strong correlation between the Draft Cyber Resilience Act (CRA) and the Cybersecurity Act [14]. European Cybersecurity Certification Schemes are applicable, and even mandatory if available for Critical Products with digital elements, for conformity assessment.

Still, European Cybersecurity Certification Schemes are required to be created to meet the specific requirements of the CRA. We note that corresponding actions are foreseen in the *Union Rolling Work Programme for European cybersecurity certification* [18].

4.6.2 Relationship of Radio Equipment Directive (RED) Cybersecurity Requirements to Cyber Resilience Act

There is an interesting overlap between Cybersecurity requirements introduced by the Radio Equipment Directive (RED) [9] for Radio Equipment and the more general requirements introduced by the Draft Cyber Resilience Act (CRA) for products with digital elements.

Specifically, RED Article 3(3)(d) relates to Cybersecurity, RED Article 3(3)(e) to Privacy and RED Article 3(3)(f) to requirements for protection from fraud. The exact text of those Articles is as follows (Table 4.1).

Table 4.1 Article 3(3) of the Radio Equipment Directive on further requirements

RED Article 3: Essential requirements
3. Radio equipment within certain categories or classes shall be so constructed that it complies with the following essential requirements: ... *(d) radio equipment does not harm the network or its functioning nor misuse network resources, thereby causing an unacceptable degradation of service;* *(e) radio equipment incorporates safeguards to ensure that the personal data and privacy of the user and of the subscriber are protected;* *(f) radio equipment supports certain features ensuring protection from fraud;* ...

The question arises on how to handle the RED related requirements in combination with the requirements introduced by the Draft Cyber Resilience Act (CRA). This issue is recognized by the Draft Cyber Resilience Act (CRA) by stating that *the essential cybersecurity requirements set out in this Regulation include all the elements of the essential requirements referred to in Article 3(3), points (d), (e) and (f) of Directive 2014/53/EU* [9]. RED Articles 3(3)(d/e/f) were originally activated through Delegated Regulation 2022/30 [19] and the Draft Cyber Resilience Act (CRA) considers to repeal or amend the Delegated Regulation [19] to resolve the overlap:

Therefore, ***when the Commission repeals or amends Delegated Regulation (EU) 2022/30*** [19] *with the consequence that it ceases to apply to certain products subject to this Regulation, the Commission and the European Standardisation Organisations should take into account the standardisation work carried out in the context of Implementing Decision C(2022) 5637* [20] *in the preparation and development of harmonised standards to facilitate the implementation of this Regulation. During the transitional period for the application of this Regulation, the Commission should provide guidance to manufacturers subject to this Regulation that are also subject to Delegated Regulation (EU) 2022/30* [19] *to facilitate the demonstration of compliance with the two Regulations.*

4.7 Derogations and Exemptions

The Draft Cyber Resilience Act (CRA) introduces a number of exemptions, especially in the field of Research and Development and Open Source.

Article 2 of the scope of Draft CRA, like in other NLF related legislations, makes some exemptions for spare parts, products linked to national security, products covered by other sector-specific legislation (automotive, medical devices). Furthermore, so-called "*free and open-source software*", defined as "*software the source code of which is openly shared and which is made available under a free and open-source license which provides for all rights to make it freely accessible, usable, modifiable and redistributable*" (Article 3), is exempted of certain requirements of the Draft CRA, for instance by being partially exempted from the reporting obligations (Article 24). Furthermore, security researchers are also listed as a group that should be exempted by the Member States of liability measures when researching vulnerabilities (Recital 75).

For conformity assessment, Article 32 of the Draft CRA includes the following exemptions:

- *Manufacturers of products with digital elements qualifying as free and open-source software* of important products with digital elements (Class I or Class II) *shall be able to demonstrate conformity with the essential cybersecurity requirements set out in Annex I by using one of the procedures referred to in paragraph 1 of this Article, provided that the technical documentation referred to in Article*

31 is made available to the public at the time of the placing on the market of those products.

- *The specific interests and needs of microenterprises and small and medium-sized enterprises, including start-ups, shall be taken into account when setting the fees for conformity assessment procedures and those fees shall be reduced proportionately to their specific interests and needs.*

4.8 Discussion of Implementation Solutions with the Object to Meet Essential Requirements of the Cyber Resilience Act

In order to meet the essential requirements set out in the Draft CRA, the authors of the present chapter propose the introduction of a number of system entities which are summarised below. It should be noted that the proposal is meant to serve as a starting point and any final product is required to pass all conformity assessment steps. The architecture is illustrated in Fig. 4.10.

User Application of the Product with Digital Elements is the functionality / service / application offered to the end user which must be complemented by entities stated in the Draft CRA in order to ensure a sufficient level of Cybersecurity.

The **Monitoring Entity** interacts with an internal and/or external database, for example a database operated remotely by the product manufacturer, in order to record internal operational data (as stated by the Draft CRA SR [3, 4]: "*providing security related information by recording and/or monitoring relevant internal activity of products with digital elements with an optout mechanism for the user*"). In case of

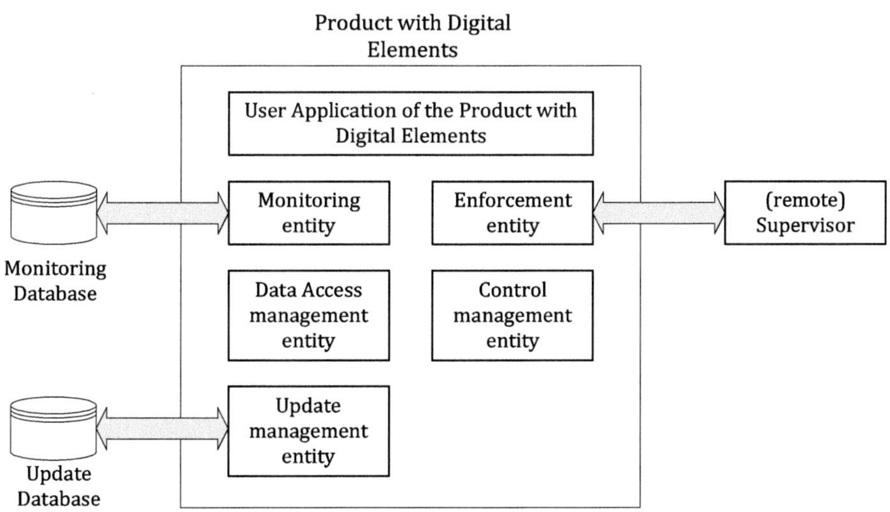

Fig. 4.10 Architecture approach meeting Draft CRA compliance requirements

a cybersecurity attack or other event, the data may be used to trace back the issue. Consequently, an update may be issued to address any detected issue / cybersecurity leak or similar.

The **Data Access Management Entity** ensures that Confidentiality of data is ensured according to the requirements of the Draft CRA SR, for example through suitable encryption: "*protecting the confidentiality of data stored, transmitted or otherwise processed by a product with digital elements*".

Integrity of private data is ensured according to the requirements of the Draft CRA SR: "*protecting the integrity of data, commands, programs by a product with digital elements, and its configuration against any manipulation or modification not authorised by the user, as well as reporting on corruptions*". The entity will request explicit authorization from the user in case that a modification of private data is requested. In case of a suspected vulnerability, the operation is interrupted and the issue is flagged to a (human) supervisor.

Access to data is enabled on a "need to know" basis, i.e. only data is accessible that is strictly needed to execute a specific task. This corresponds to the following requirement of the Draft CRA SR: "*processing only personal or other data that are adequate, relevant and limited to what is necessary in relation to the intended purpose of the product with digital elements ('minimisation of data')*".

The **Control Management Entity** ensures that only authorized users will be able to access the product with digital elements. For this purpose, the entity includes authentication, identity or access management systems. Also, the entity manages various levels of access depending on the users, for example certain users may only have limited access to some applications, other users may have no access at all and a superuser may have access even to engineering/supervision functionalities and similar. All related features are provided as required by the Draft CRA SR: "*protection of products with digital elements from unauthorised access and reporting on possible unauthorised access*".

The **Update Management Entity** constantly interacts with an external Updates Database that maintains (critical) updates in order to mitigate eventual vulnerabilities. In case that the target device detects a new (suspected) vulnerability, it is reported to the Database or to any other suitable (government) agency. The requirement of the Draft CRA SR is stated as follows: "*vulnerabilities in products with digital elements can be addressed through security updates*".

4.9 Conclusions

The Cyber Resilience Act (CRA) [1, 2, 5] will be a major European Regulation governing access to the European Union Single Market for *products with digital elements*, which will likely apply to (almost) any digital finished product or component. In addition to requirements for placement of products onto the market, the CRA will additionally introduce requirements applying over the lifetime of the product (no less than 5 years unless product life time is shorter), especially related to vulnerabilities management. The CRA will significantly change market access mechanisms for stakeholders, thus it is important to understand the new requirements and to implement relevant processes such that the access to the European Union Single Market is maintained for the future.

This chapter details the content of the regulation and its on economic operators. A dedicated sections proposes a possible implementation solution with the object to meet essential requirements of the Cyber Resilience Act and should be enhanced in the future according to the experience of manufacturers.

Disclaimer: Note that the authors have assembled the information above to the best of their knowledge as available at the time when the present book was written. Any interpretation is the personal opinion of the authors. Also, the legislation framework can change and evolve any time. The only legally binding text is the one provided in the documents formally published in the European Union Official Journal.

References

1. Proposal for a REGULATION OF THE EUROPEAN PARLIAMENT AND OF THE COUNCIL on horizontal cybersecurity requirements for products with digital elements and amending Regulation (EU) 2019/1020, 15th September 2022, available at https://eur-lex.europa.eu/legal-content/EN/TXT/?uri=celex%3A52022PC0454
2. Draft Cyber Resilience Act, Text agreed during interinstitutional negotiations, December 2023, available at https://oeil.secure.europarl.europa.eu/oeil/popups/ficheprocedure.do?lang=en&reference=2022/0272(COD), https://www.europarl.europa.eu/RegData/commissions/itre/inag/2023/12-20/ITRE_AG(2023)758004_EN.docx
3. Draft standardisation request to European Standards Organisations in support of Union policy on cybersecurity requirements for products with digital elements, 17th April 2024, available at https://ec.europa.eu/docsroom/documents/58974
4. Draft COMMISSION IMPLEMENTING DECISION of XXX on a standardisation request to the European Committee for Standardisation (CEN), the European Committee for Electrotechnical Standardisation (Cenelec) and the European Telecommunications Standards Institute (ETSI) as regards products with digital elements in support of Regulation [...], 20th September 2024
5. REGULATION OF THE EUROPEAN PARLIAMENT AND OF THE COUNCIL on horizontal cybersecurity requirements for products with digital elements and amending Regulations (EU) No 168/2013 and (EU) 2019/1020 and Directive (EU) 2020/1828 (Cyber Resilience Act), 25 September 2024, version adopted by the European Council on 10 October 2024, available at https://data.consilium.europa.eu/doc/document/PE-100-2023-INIT/en/pdf
6. Cyber resilience act: Council adopts new law on security requirements for digital products, Press Release, Council of the European Union, 10th October 2024, available at https://www.consilium.europa.eu/en/press/press-releases/2024/10/10/cyber-resilience-act-council-adopts-new-law-on-security-requirements-for-digital-products/
7. Chiara, P.G. The Cyber Resilience Act: the EU Commission's proposal for a horizontal regulation on cybersecurity for products with digital elements. Int. Cybersecur. Law Rev. 3, 255–272 (2022). https://doi.org/10.1365/s43439-022-00067-6

8. Meagher, H., Dhirani, L.L. (2024). Cyber-Resilience, Principles, and Practices. In: Naseer Qureshi, K., Newe, T., Jeon, G., Chehri, A. (eds) Cybersecurity Vigilance and Security Engineering of Internet of Everything. Internet of Things. Springer, Cham. https://doi.org/10.1007/978-3-031-45162-1_4
9. DIRECTIVE 2014/53/EU OF THE EUROPEAN PARLIAMENT AND OF THE COUNCIL of 16 April 2014 on the harmonisation of the laws of the Member States relating to the making available on the market of radio equipment and repealing Directive 1999/5/EC
10. Commission Notice — The 'Blue Guide' on the implementation of EU products rules 2016 (Text with EEA relevance), available at https://eur-lex.europa.eu/legal-content/EN/TXT/?uri=CELEX%3A52016XC0726%2802%29
11. Commission Delegated Regulation (EU) 2022/30 of 29 October 2021 supplementing Directive 2014/53/EU of the European Parliament and of the Council with regard to the application of the essential requirements referred to in Article 3(3), points (d), (e) and (f), of that Directive (Text with EEA relevance)
12. Cyber Resilience Act European Commission, DG CONNECT, presentation to Cybersecurity and Infrastructure Security Agency (CISA), United States, available at https://www.cisa.gov/sites/default/files/2023-09/EU%20Commission%20SBOM%20Work_508c.pdf
13. NEW EU CYBER RESILIENCE ACT TO HAVE GLOBAL IMPACT…SOON, FM Approvals, available at https://www.fmapprovals.com/Product-Alerts-and-News-Events/Insights/eu-cyber-resilience-act
14. Regulation (EU) 2019/881 of the European Parliament and of the Council of 17 April 2019 on ENISA (the European Union Agency for Cybersecurity) and on information and communications technology cybersecurity certification and repealing Regulation (EU) No 526/2013 (Cybersecurity Act) (Text with EEA relevance), available at https://eur-lex.europa.eu/legal-content/EN/TXT/?uri=celex%3A32019R0881
15. The Vienna Agreement, CEN/CENELEC, available at https://www.cencenelec.eu/about-cen/cen-and-iso-cooperation/
16. The Frankfurt Agreement, CEN/CENELEC, available at https://www.cencenelec.eu/about-cenelec/cenelec-and-iec-cooperation/
17. Cyber Resilience Act Requirements Standards Mapping - Joint Research Centre & ENISA Joint Analysis, April 2024, available at https://www.enisa.europa.eu/publications/cyber-resilience-act-requirements-standards-mapping
18. COMMISSION STAFF WORKING DOCUMENT: Union Rolling Work Programme for European cybersecurity certification, February 2024, available at https://digital-strategy.ec.europa.eu/en/library/union-rolling-work-programme-european-cybersecurity-certification
19. COMMISSION DELEGATED REGULATION (EU) 2022/30 of 29 October 2021 supplementing Directive 2014/53/EU of the European Parliament and of the Council with regard to the application of the essential requirements referred to in Article 3(3), points (d), (e) and (f), of that Directive (Text with EEA relevance), available at http://data.europa.eu/eli/reg_del/2022/30/2023-10-27
20. COMMISSION IMPLEMENTING DECISION on a standardisation request to the European Committee for Standardisation and the European Committee for Electrotechnical Standardisation as regards radio equipment in support of Directive 2014/53/EU of the European Parliament and of the Council and Commission Delegated Regulation (EU) 2022/30, available at https://ec.europa.eu/transparency/documents-register/detail?ref=C(2022)5637&lang=en

Chapter 5
Introduction to the European Radio Equipment Directive

Markus Mueck, John Roman, Christophe Gaie, Stéphane Du Boispéan, Ahmad Al Shamsi, Sultan Albalooshi, Mohammad Alshamsi, Shaikha Almulla, Lina Bariah, and Merouane Debbah

Abstract The European Radio Equipment Directive (RED) *establishes a regulatory framework for placing radio equipment on the market*. RED thus governs access to the European Union Single Market and introduces so-called *essential requirements* that need to be complied with by any product under the scope of RED. Originally,

---

M. Mueck (✉)
Intel Deutschland GmbH, Munich, Germany
e-mail: Markus.Dominik.Mueck@intel.com

J. Roman
Intel Corp., Santa Clara, USA
e-mail: John.M.Roman@intel.com

C. Gaie
Head of the Engineering and Digital Innovation Division, Prime Minister Services, Paris, France
e-mail: christophe.gaie@gmail.com

S. Du Boispéan
Intel Corporation SA/NV, Kontich, Belgium
e-mail: Stephane.du.Boispean@intel.com

A. Al Shamsi · S. Albalooshi · M. Alshamsi · S. Almulla
Telecommunications and Digital Government Regulatory Authority (TDRA), Abu Dhabi, United Arab Emirates
e-mail: Ahmad.AlShamsi@tdra.gov.ae

S. Albalooshi
e-mail: Sultan.Albalooshi@tdra.gov.ae

M. Alshamsi
e-mail: Mohammad.AlShamsi@tdra.gov.ae

S. Almulla
e-mail: Shaikha.AlMulla@tdra.gov.ae

L. Bariah · M. Debbah
Khalifa University, Abu Dhabi, United Arab Emirates
e-mail: Lina.Bariah@ku.ac.ae

M. Debbah
e-mail: Merouane.Debbah@ku.ac.ae

M. Mueck and C. Gaie (eds.), *European Digital Regulations*, Intelligent Systems Reference Library 265, https://doi.org/10.1007/978-3-031-80809-8_5

those essential requirements were limited to (i) *the protection of health and safety of persons and of domestic animals and the protection of property*, (ii) *electromagnetic compatibility* and (iii) *the efficient and effective use of spectrum.* More recently, additional requirements were put into force through Delegated Acts which were beforehand inactive. Those include requirements related to introducing a common charger, ensuring caller location in emergency communications from mobile devices as well as requirements on Cyber Security and Privacy. Furthermore, the possible activation of Article 3(3)(i) and Article 4 are under discussion with the objective to ensure that the combination of software and hardware remains compliant after installation of software updates. In the present chapter, we give an overview of all those elements and further outline the expected relation of RED to the future Cyber Resilience Act.

Keywords Regulation · Radio equipment · Spectrum · Software reconfiguration · Cyber security · Privacy

5.1 Introduction

As stated on the European Commission Website [1], the European Radio Equipment Directive (RED) [2] creates a regulatory framework for the market introduction of radio equipment. It is thus essential for all manufacturers to understand whether (i) a specific product, component, etc. is in scope of RED and (ii) which essential requirements need to be fulfilled in such a case. As it is explained in the RED guide [3], not only finished products but also certain Radio Frequency (RF) components and certain Integrated Circuits (ICs), etc. are in scope of RED and thus need to meet the essential requirements as stand-alone product before being placed on the European Union Single Market. Recently, two amendments have been published introducing a *common charging* solution: Directive (EU) 2022/2380 [4] and Commission Delegated Regulation (EU) 2023/1717 [5].

Article 3 of RED contains the *essential requirements* which need to be met by manufacturers to access the European Union Single Market. With RED being a regulation under the New Legislative Framework (NLF), European Standardisation Organisations (ESOs), ETSI, CEN and CENELEC, were tasked to develop related Harmonised Standards which provide presumption of conformity under the condition that they are published in the EU Official Journal. Standardisation Request M/536 [6] provides the corresponding legal framework.

Harmonised Standards on electromagnetic compatibility requirements, as laid out in Article 3.1, are provided by ETSI's EN 301 489 series. ETSI EN 301 489 Part 1 [7] covers common technical requirements applicable to all radio equipment, while ETSI EN 301 489 Parts 2–54 address specific cases (Table 5.1).

Article 3.2 relates to the efficient and effective use of spectrum. Additional details are provided in ETSI Guide EG 203 367 [8] including guidance on RF requirements to be included into Harmonised Standards, such as radiated emissions, conducted

Table 5.1 Article 3(1) of the radio equipment directive on protection of health and safety and electromagnetic compatibility

RED Article 3: Essential requirements
1. Radio equipment shall be constructed so as to ensure: *(a) the protection of health and safety of persons and of domestic animals and the protection of property, including the objectives with respect to safety requirements set out in Directive 2014/35/EU, but with no voltage limit applying;* *(b) an adequate level of electromagnetic compatibility as set out in Directive 2014/30/EU.*

emissions, harmonic current emissions, etc. Also, this ETSI Guide addresses the case of '*combined equipment*', for example when integrating a radio product (such as a RLAN product or other) into a non-radio product (such as a refrigerator or other).

Contrary to Article 3.1 of RED, there is no single series of Harmonised Standards available for Article 3.2. Rather, ETSI has published independent Harmonised Standards for specific frequency bands and systems. For example, ETSI EN 301 893 [9] is the RED Article 3.2 Harmonised Standard for 5 GHz Radio Local Area Networks (RLAN) (Table 5.2).

Article 3.3 includes further essential requirements, but they have not originally been in force when RED was published in 2014. The following sub-articles of Article 3.3 have been activated so far:

- Article 3(3)(a) on the common charger: As stated by the European Commission [1], *USB-C will be the common port. This will allow consumers to charge their devices with any USB-C charger, regardless of the device brand.* Directive (EU) 2022/2380 [4] defines the requirements of the 'common charging' solution and Commission Delegated Regulation (EU) 2023/1717 [5] updates the references to technical specifications for wired charging.
- Article 3(3)(d), Article 3(3)(e) and Article 3(3)(f) relate to requirements on Cyber Security and Privacy as further detailed in the delegated regulation 2022/30 [10]. A corresponding Standardisation Request was issued to CEN and CENELEC [11] including an amendment [12]. These Articles will be discussed below in further detail.
- Article 3(3)(g) ensures caller location in emergency communications from mobile devices, relying on Global navigation satellite system (GNSS), and specifically the European Galileo system, and WiFi. The Article was activated through Delegated Regulation 2019/320 [13]. The RED Guide [3] lists references [14–18] for a corresponding implementation.
- Article 3(3)(i) and Article 4 relate to Software Reconfigurability with the objective to maintain compliance of the combination of software and hardware in case of software upload. There has been extensive discussion on these two articles in European Commission Expert Group RRS (Reconfigurable Radio Systems) and later Expert Group RE. No related Delegated Act has been issued yet, but the ETSI Technical Committee Reconfigurable Radio Systems (TC RRS) has been working extensively on solutions possibly supporting an implementation of RED Article 3(3)(i) and Article 4. Those will be discussed below in further detail (Table 5.3).

Furthermore, in cases of restrictions on putting into service or of requirements for authorisation of use, Implementing Regulation 2017/1354 [19] specifies how to present the information provided for in Article 10(10) of RED.

It is noteworthy that the European Commission has created an Expert Group Radio Equipment (EG RE) [20] where new initiatives and open questions are being discussed with relevant stakeholders.

Table 5.2 Article 3(2) of the radio equipment directive on effective and efficient usage of spectrum

RED Article 3: Essential requirements
2. Radio equipment shall be so constructed that it both effectively uses and supports the efficient use of radio spectrum in order to avoid harmful interference.

Table 5.3 Article 3(3) of the radio equipment directive on further requirements

RED Article 3: Essential requirements
3. Radio equipment within certain categories or classes shall be so constructed that it complies with the following essential requirements: *(a) radio equipment interworks with accessories, in particular with common chargers;* *(b) radio equipment interworks via networks with other radio equipment;* *(c) radio equipment can be connected to interfaces of the appropriate type throughout the Union;* *(d) radio equipment does not harm the network or its functioning nor misuse network resources, thereby causing an unacceptable degradation of service;* *(e) radio equipment incorporates safeguards to ensure that the personal data and privacy of the user and of the subscriber are protected;* *(f) radio equipment supports certain features ensuring protection from fraud;* *(g) radio equipment supports certain features ensuring access to emergency services;* *(h) radio equipment supports certain features in order to facilitate its use by users with a disability;* *(i) radio equipment supports certain features in order to ensure that software can only be loaded into the radio equipment where the compliance of the combination of the radio equipment and software has been demonstrated.* *The Commission shall be empowered to adopt delegated acts in accordance with Article 44 specifying which categories or classes of radio equipment are concerned by each of the requirements set out in points (a) to (i) of the first subparagraph of this paragraph.*

5.2 Requirements Related to Cyber Security and Privacy

Cyber Security and Privacy related requirements are introduced through the activation of RED Article 3(3)(d), Article 3(3)(e) and Article 3(3)(f) as further detailed in the Delegated Regulation 2022/30 [10] and the corresponding Standardisation Request [11] including an amendment [12]. This Standardisation Request was specifically addressed to ESOs CEN/CENELEC which led to the creation of EN 18031-1:2024 [21] in support of RED Article 3(3)(d), EN 18031-2:2024 [22] in support of RED Article 3(3)(e) and EN 18031-3:2024 [23] in support of RED Article 3(3)(f). At the time when the present chapter was written, those ENs have not been listed in the European Union Official Journal and it remains to be seen to which extent the European Commission will authorize their usage for presumption of conformity with RED Articles (3)(3)(d/e/f).

The overall sequence of actions is illustrated below (Fig. 5.1):

For European Standards Organisations (ESOs), it was a specific challenge to develop Harmonised Standards [21–23] in support of RED Article 3(3)(d), Article 3(3)(e) and Article 3(3)(f). The main reason is that the type of requirements is fundamentally different compared to those introduced by previously applicable Article 3(1) and Article 3(2). The requirements introduced by the latter Articles relate to physically observable requirements, such as limitations on output power levels, out of band radiation levels, etc. Those can indeed be measured and verified, if needed several times with suitable equipment. I.e., the corresponding requirements are *objectively and independently verifiable* and rely on *verifiable and reproducible approaches* as

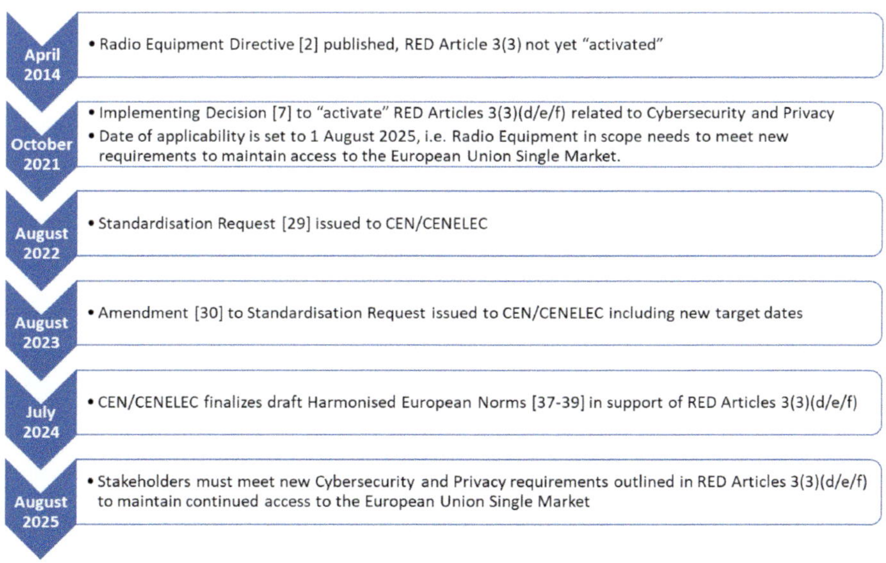

Fig. 5.1 Time-Line for implementation of RED Articles 3(3)(d/e/f) on cybersecurity and privacy

required by [11]. On the other hand, requirements related to Cyber Security and Privacy as introduced by RED Article 3(3)(d), Article 3(3)(e) and Article 3(3)(f) are rather of functional nature, such as requirements on user authentication and authorization and similar. Compliance to such requirements cannot be directly measured and thus other ways of conformity assessment need to be defined. This is a paradigm shift for conformity assessment, because it is challenging to define appropriate *test methods or equivalent verifiable and reproducible approaches* [11]. The idea is that two independent test houses should come to the same conclusions for a given product. This objective is traditionally achieved for physically observable requirements, but it is far more challenging for functional requirements. To the best knowledge of the authors of the present chapter, no fully satisfactory approach is available as of now.

Also, it may defy the purpose to define specific Cyber Security related solutions in a Harmonised Standard in too great detail. Indeed, this information would likely be exploited by a malicious attacker to circumvent the protection mechanisms.

The exact scope of applicability of RED Article 3(3)(d) is defined in Article 1(1) of [10], of RED Article 3(3)(e) in Article 1(2) of [10] and of RED Article 3(3)(f) in Article 1(3) of [10]. Article 1(1) of [10] is as follows:

The essential requirement set out in Article 3(3), point (d), of Directive 2014/53/EU shall apply to any radio equipment that can communicate itself over the internet, whether it communicates directly or via *any other equipment ('internet-connected radio equipment').*

Also, the related Standardisation Request [11] indicates that

Commission Delegated Regulation (EU) 2022/303 specifies that the essential requirement set out in Article 3(3), point (d), of Directive 2014/53/EU is applicable to internet-connected radio equipment.

To the best knowledge of the authors of the present chapter, it is not entirely clear which type of equipment falls or does not fall under the above stated scope. Indeed, any source of data—if used in combination with some radio equipment—may be linked directly or indirectly to an internet-connection. For example, a Bluetooth headset is likely not a directly internet-connected device, but when connected to a Smartphone or similar, an indirect internet connection is achieved and spoken language may be distributed over the internet. Thus, does such a Bluetooth headset fall under the scope of RED Article 3(3)(d)? Further guidance on the exact applicability of the regulation is needed.

If the European Commission decides to list the Harmonised European Norms EN 18031-1:2024 [21] in support of RED Article 3(3)(d), EN 18031-2:2024 [22] in support of RED Article 3(3)(e) and EN 18031-3:2024 [23] in support of RED Article 3(3)(f), then stakeholders can use those for self-declaration of conformity to the essential requirements of RED Articles 3(3)(d/e/f). Alternatively, stakeholders may seek the help of accredited conformity assessment bodies (Notified Bodies under the RED) which are authorized to provide an assessment of conformity to the essential requirements of RED. Without the availability of Harmonised European Norms, only the latter solution through Notified Bodies remains available to maintain continued access to the European Union Single Market for products under scope of RED Articles 3(3)(d/e/f).

Furthermore, we note that the Delegated Regulation 2022/30 [10] introduces a number of exemptions for specific cases where the requirements are already covered through other existing Union legislation (Table 5.4):

Article 2(1) of the Delegated Regulation 2022/30 [10] thus introduces exemptions for all Cybersecurity and Privacy related Articles 3(3)(d/e/f) of RED for radio equipment related to the following:

- Medical Devices covered by Regulation (EU) 2017/745 [24];
- In vitro diagnostic medical devices covered by Regulation (EU) 2017/746 [25].

Furthermore, Article 2(2) of the Delegated Regulation 2022/30 [10] introduces exemptions for Cybersecurity and Privacy related Articles 3(3)(e/f) of RED for radio equipment related to the following (i.e., Article 3(3)(d) still applies to the concerned radio equipment):

- Civil aviation covered by Regulation (EU) 2018/1139 [26];
- Motor vehicles and their trailers, and systems, components and separate technical units intended for such vehicles, as regards their general safety and the protection of vehicle occupants and vulnerable road users covered by Regulation (EU) 2019/2144 [27];
- Electronic road toll systems and facilitating cross-border exchange of information on the failure to pay road fees in the Union covered by Directive (EU) 2019/520 [28].

5.3 Requirements Related to Software Reconfiguration as Developed by ETSI's Technical Committee on Reconfigurable Radio Systems (TC RRS)

RED Article 3(3)(i) introduce requirements related to the combination of software and hardware, ensuring a continued compliance to the essential requirements in case that a software update is installed (Table 5.5):

Also, we note that RED Article 4 is closely related to Software Reconfiguration and needs to be considered for the possible future activation of RED Article 3(3)(i) (Table 5.6):

This topic has been extensively discussed in European Commission Expert Group RRS (Reconfigurable Radio System) and European Commission Expert Group RE (Radio Equipment). At the time when the present chapter was crafted, a potential activation of RED Article 3(3)(i), together with related Article 4 through a Delegated Act was being discussed in EC Expert Group RE [20]. The situation needs to be further monitored for the next steps.

We note that the solutions developed by ETSI TC RRS may equally apply in support RED Articles 3(3)(d/e/f) as well as in support of the Cyber Resilience Act (CRA) where secure software updates over the lifetime of a product are one of the key objectives.

Table 5.4 Article 2 of the Delegated Regulation 2022/30 related to the activation of cybersecurity and privacy related Articles 3(3)(d/e/f) of the radio equipment directive

Article 2

1. *By way of derogation from Article 1, the essential requirements set out in Article 3(3), points (d), (e) and (f), of Directive 2014/53/EU shall not apply to radio equipment to which either of the following Union legislation also applies:*

   (a) Regulation (EU) 2017/745;
   (b) Regulation (EU) 2017/746.

▼

2. *By way of derogation from Article 1(2) and Article 1(3), the essential requirements set out in Article 3(3), points (e) and (f), of Directive 2014/53/EU shall not apply to radio equipment to which any of the following Union legislation also applies:*

   (a) Regulation (EU) 2018/1139;
   (b) Regulation (EU) 2019/2144;
   (c) Directive (EU) 2019/520

Table 5.5 Article 3(3)(i) of the radio equipment directive on software reconfiguration on radio equipment

RED Article 3: Essential requirements
…
3. Radio equipment within certain categories or classes shall be so constructed that it complies with the following essential requirements:
…
(i) radio equipment supports certain features in order to ensure that software can only be loaded into the radio equipment where the compliance of the combination of the radio equipment and software has been demonstrated.
…

Table 5.6 Article 4 of the radio equipment directive on software reconfiguration on radio equipment

Article 4 *Provision of information on the compliance of combinations of radio equipment and software*
1. *Manufacturers of radio equipment and of software allowing radio equipment to be used as intended shall provide ... information on the compliance of intended combinations of radio equipment and software with the essential requirements set out in Article 3. ..., the information shall precisely identify the radio equipment and the software which have been assessed, and it shall be continuously updated.* 2. *The Commission shall be empowered to adopt delegated acts ... specifying which categories or classes of radio equipment are concerned ....* 3. *The Commission shall adopt implementing acts laying down the operational rules for making the information on compliance available for the categories and classes specified by the delegated acts adopted ....*

ETSI Technical Committee Reconfigurable Radio Systems (ETSI TC RRS) has been extensively studying such systems and has published an extensive list of specifications [29–44] as well as a summary White Paper [45].

ETSI EN 303 648 [30] introduces an overall architecture of a Reconfigurable Radio System with a Communication Service Layer (CSL) coordinating the operations of one or multiple Radio Computers as illustrated in Fig. 5.2.

In this context, ETSI TC RRS has introduced a number of interfaces—including the "generalized MUlti-Radio Interface (gMURI)" as illustrated in Fig. 5.3.

ETSI EN 303 303 681-1 [31] summarizes available services as defined for gMURI by ETSI [31]. In this context, ETSI RC RRS uses the terminology "*RadioApp*" for any software component affecting the radio characteristics of a Radio Equipment. This distinction is important, since there is no intention to regulate general software, such as Video Games, Word processing, etc. Rather, the intention of the Radio Equipment Directive and especially Article 3(3)(i) is to focus on Software that may alter the radio characteristics and thus put the combination of Software and Hardware into a non-confirming state (Fig. 5.4).

In addition to the upper features, a security analysis has been conducted, the results are available in ETSI TR 103 087 [42] and ETSI TS 103 436 [43]. The following security measures are introduced to address these threats in the ETSI Software Reconfiguration Framework:

- Solutions ensuring the integrity of the Radio Applications, RE Configuration Policy and Declaration of Conformity;
- Solutions verifying the identity of the developer of Radio Applications, the issuer of the RE Configuration Policy, and the issuer of the Declaration of Conformity;
- Solutions preventing an asset installation when the asset is not provided by a legitimate actor;
- Solutions enabling the use of the reconfiguration feature as a security update mechanism;
- Solutions for proving conformance of the radio platform and radio application to the regulatory Declaration of Conformity, considering that the set of installed radio applications can change over time;
- Solutions for preventing illegitimate use of the Declaration of Conformity (in particular against counterfeit);
- Solutions for providing audit functionalities including a non-repudiation framework and remote attestation;
- Solutions enabling a long-term management framework (e.g., transition of equipment responsibility from one manufacturer to another);
- Solutions for preventing masquerade of stakeholders in the RRS value chain;
- Prevention of code theft; and,
- Solutions to support supply chain integrity and assurance (which underpins all of the above measures).

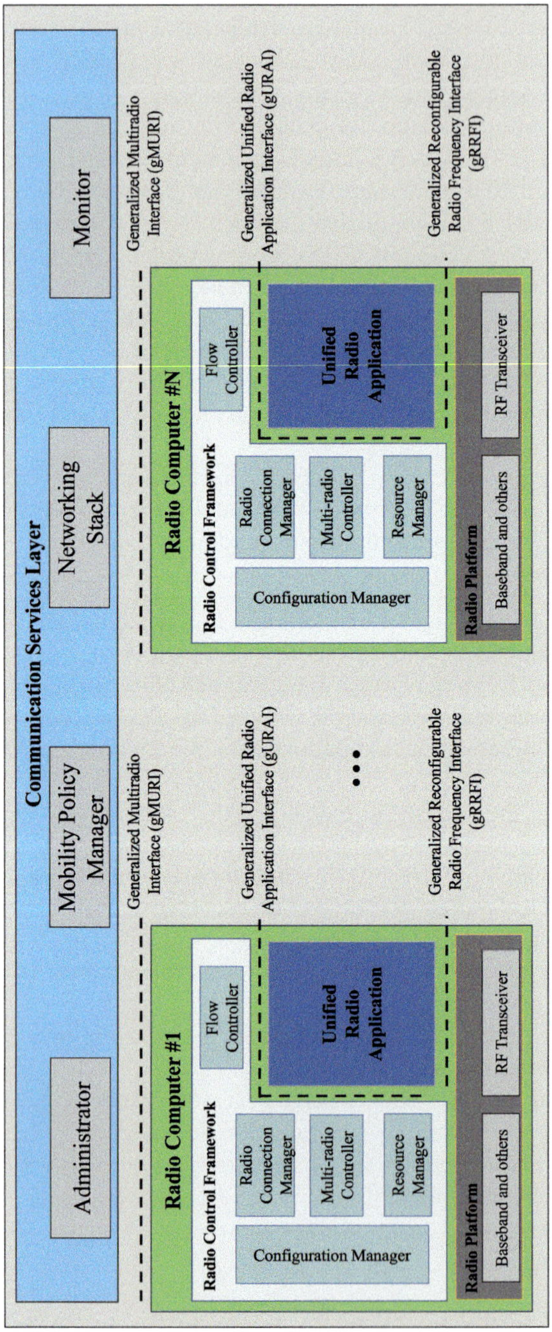

Fig. 5.2 Reconfigurable radio equipment architecture components for radio reconfiguration as defined in ETSI EN 303 648 [30]

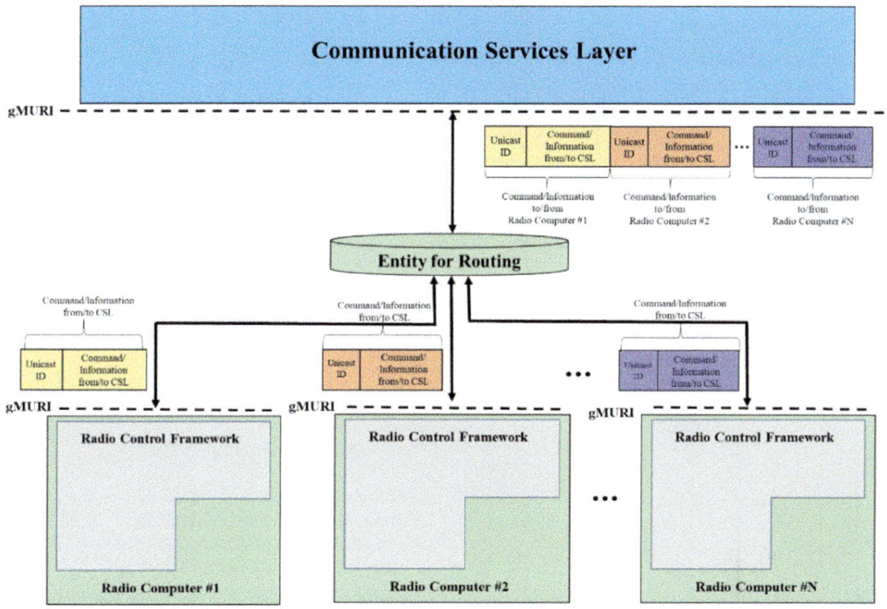

Fig. 5.3 ETSI RRS generalized multi-radio interface (gMURI) as defined in ETSI EN 303 648 [30]

5.4 Possible Approach for a Future Harmonised Standard in Support of RED Article 3(3)(I) on Software Reconfiguration

In case that RED Article 3(3)(i), possibly in combination with RED Article (4), will be activated through a Delegated Act in the future, a Harmonised Standard in support of the Article can be created based on the available building blocks outlined in the section above.

Indeed, any relevant Software Components, also called *RadioApps* or *Unified Radio Applications (URA)* in ETSI TC RRS terminology, affecting the radio characteristics of a concerned Radio Equipment, may be carried by a so-called *Radio Application Package (RAP)* as defined in ETSI TS 103 850 [44]. The overall structure is summarized below (Fig. 5.5):

The content of the top level structure of the Radio Application Package is further detailed below (Table 5.7).

With the upper approach, any software update is first tested, verified and validated by responsible entity (typically the manufacturer of the Radio Equipment or other). Once the combination of the available Hardware and the novel software component(s) are leading to a fully compliant Radio Equipment, then the new software components are delivered through a Radio Application Package together with additional information on Security, etc. to the target Radio Equipment. The target Radio

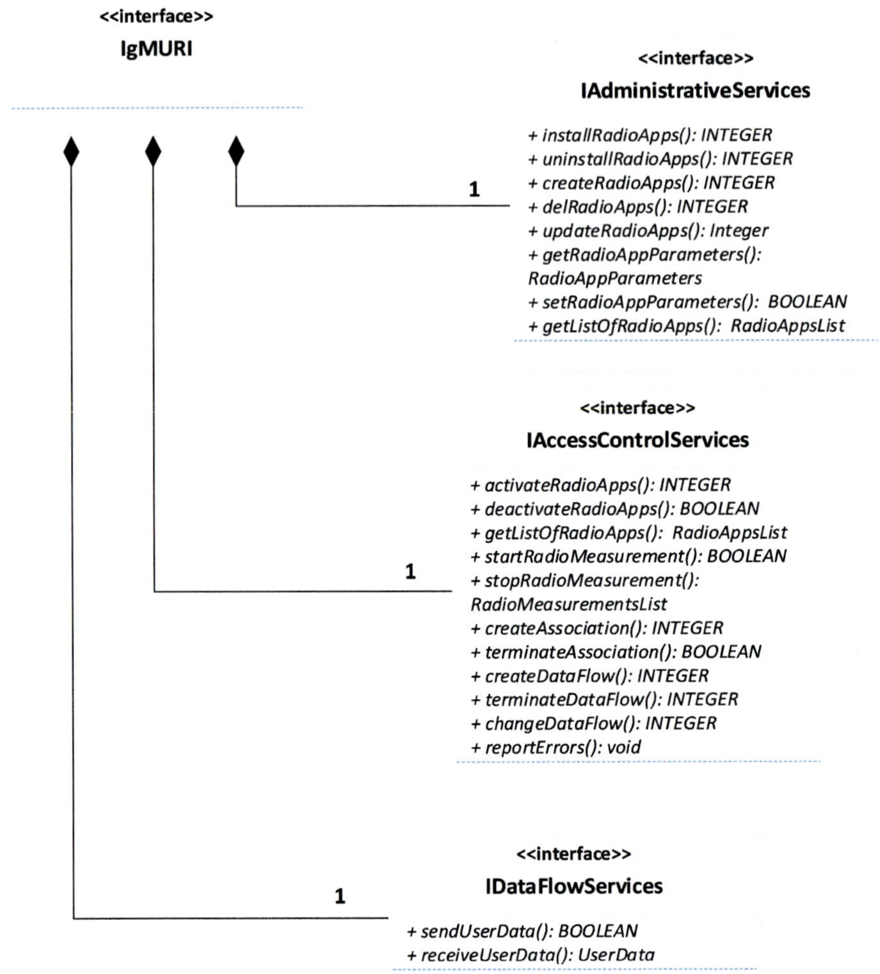

Fig. 5.4 Services introduced by the ETSI RRS generalized multi-radio interface (gMURI) as defined in ETSI EN 303 681-1 [31]

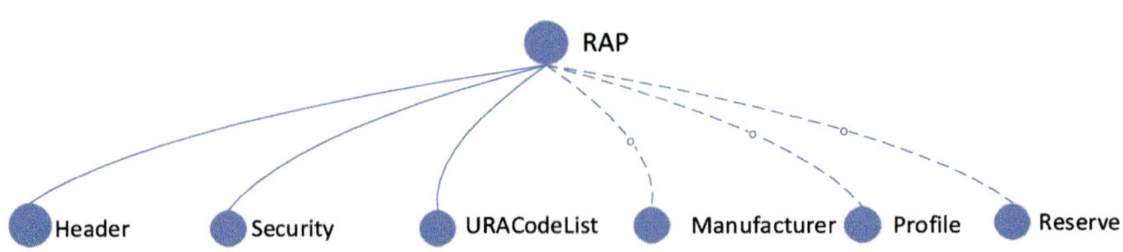

Fig. 5.5 Top level structure of the radio application package as defined in ETSI TS 103 850 [44]

Table 5.7 Content of the radio application package as defined in ETSI TS 103 850 [44]

Item number	Node name	Node type	Bit field
1	Header	Mandatory	The RAP header
2	Security	Mandatory	The security section
3	URACodeList	Mandatory	Unified radio application (URA) codes (any representation). The URA code list consisting of code sections for a few radio applications
4	Manufacturer	Optional	The manufacturer information section
5	Profile	Optional	The initial profile section
6	Reserve	Optional	Reserved for the future use

Equipment can then use information contained in the Radio Application Package to verify whether the software is originating from an authorized source, whether the integrity of the software is confirmed (i.e., so changes were applied during the delivery process), etc. Only when all verification steps are completed in the target Radio Equipment, then the software is accepted, installed and executed.

Finally, ETSI has provided an input document [46] to EC Expert Group Radio Equipment offering details on how a verification of a target Radio Equipment may be conducted. Indeed, ETSI considers as one approach to provide a "Test Access" interface, which enables measurement equipment to directly access to the relevant interfaces to be verified. The basic principle is illustrated in Fig. 5.6:

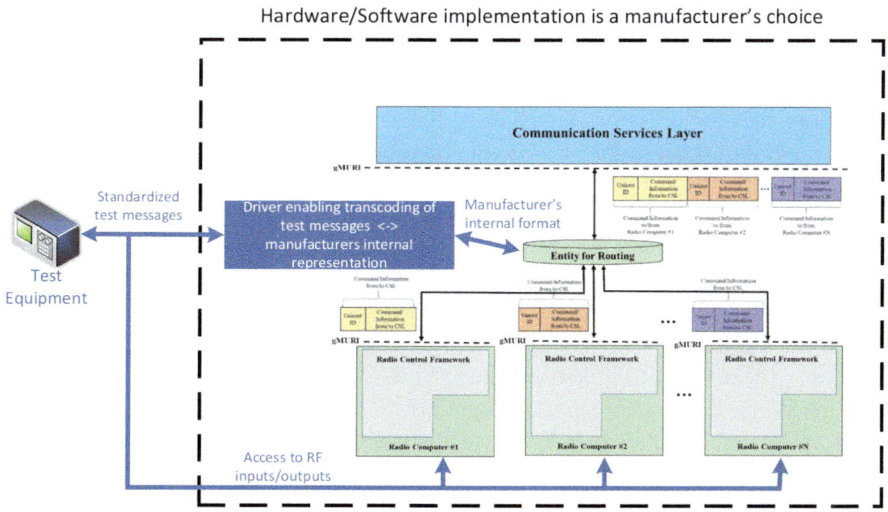

Fig. 5.6 The equipment under test supports a "Test Access" which grants direct access to internal interfaces [46]

With the approach illustrated in Fig. 5.6, test equipment would be able to use standardized test messages to invoke and verify all services defined for internal interfaces [29–44] of the ETSI Software Reconfiguration Architecture. A manufacturer would be required to implement a Device Driver which transcodes the standardized test messages into any applicable internal format according to manufacturers' choices.

5.5 Interdependencies with Other European Regulations

As explained above, Cyber Security and Privacy related requirements are introduced through the activation of RED Article 3(3)(d), Article 3(3)(e) and Article 3(3)(f). These RED Articles were activated through Delegated Regulation 2022/30 [10] followed by the European Commission issuing the corresponding Standardisation Request [11] including an amendment [12]. We note that there is an almost complete overlap with the future Cyber Resilience Act (CRA) which is available as a Draft [47] at the time when the present chapter was written. The present book contains a stand-alone chapter providing a comprehensive introduction to the CRA.

This issue is recognized by the Draft Cyber Resilience Act (CRA) by stating that *the essential cybersecurity requirements set out in this Regulation include all the elements of the essential requirements referred to in Article 3(3), points (d), (e) and (f) of Directive 2014/53/EU* [2]. The Draft Cyber Resilience Act (CRA) considers to repeal or amend the Delegated Regulation [10] to resolve the overlap:

Therefore, in case that the Commission repeals or amends Delegated Regulation (EU) 2022/30 [10], affecting certain products covered by this Regulation, the Commission and European Standardisation Organisations should consider the standardization work done under Implementing Decision C(2022) 5637 [11]. This will help them create new harmonised standards to support the implementation of this Regulation.

During the transitional period for this Regulation, the Commission should guide manufacturers who are subject to both this Regulation and Delegated Regulation (EU) 2022/30 [10] to help them demonstrate compliance with both regulations.

As discussed above, the related Standardisation Request was specifically addressed to ESOs CEN/CENELEC which led to the creation of EN 18031-1:2024 [21] in support of RED Article 3(3)(d), EN 18031-2:2024 [22] in support of RED Article 3(3)(e) and EN 18031-3:2024 [23] in support of RED Article 3(3)(f). It is therefore expected that the corresponding standardisation deliverables will be moved from the Radio Equipment Directive under the Cyber Resilience Act. However, the scope of the Cyber Resilience Act is substantially broader (addressing any product with digital elements, i.e. any software or hardware product and its remote data processing solutions), so the appropriate applicability of the available ENs needs to be monitored.

5.6 Comparison of Radio Regulation Between European and the United States

While the present chapter mainly focuses on the Radio Equipment Regulation framework in Europe, we believe that it is important to give a perspective on how the corresponding approaches differ in other regions. Below is a summary on the regulation framework on Radio Equipment is operating in the United States as governed by the Federal Communications Commission (FCC).

The Federal Communications Commission (FCC) regulates the operation, use and sale of radio equipment. Established in 1934, they also regulate the commercial spectrum assignment and use, as well as broadcasting. The FCC is organized into six operating Bureaus and ten Staff Offices, whose role is to provide rules and process for compliance. The Office of Engineering and Technology contains divisions for Radio Policy and Rules, Electromagnetic Compatibility (EMC), and the laboratory division manages the procedures and process for equipment compliance.

The radio regulations are provided in the Code of Federal regulations (CFR), Title 47: telecommunications. The regulations are broken into parts, where the rules for operation of wireless equipment are provided in Part 15: Radio Frequency Devices. Part 15 covers all types of radiators, whether incidental, unintentional, or intentional with technical rules for operation to address interference. Intentional radiators are broken down further into licensed and unlicensed, where unlicensed devices must not cause and also accept interference. Licensed radiators, for example cell phones and networks, are afforded protection from interference.

New rules, or changes in the existing rules are handled through a structured rule making process, where for example a Notice of Proposed Rule Making, or NPRM is issued. In some cases, the FCC may first issue a notice of inquiry (NOI). These are then published in the US Federal Register, which starts the comment clock, typically 30–60 days. Comments are accepted from all parties, and typically a reply comment period is also provided, where additional comments can be provided.

For radio type approval, which the FCC calls equipment authorization, the laboratory division, provides guidance for demonstrating compliance to the rules via the Knowledge Database (KDBs). Manufacturers may also make a confidential inquiry to the FCC via the KDB process, to obtain an interpretation of the requirements for their specific case.

The FCC has also created a program to allow independent reviewers qualify as Telecommunication Certification Bodies (TCBs), where they can review the test results and equipment authorization packages from accredited testing laboratories, and provide the approval for some types of radio equipment. For new and complex technologies, the FCC has the Pre Approval Guidance (PAG), where the TCBs/ manufacturers must work directly with the labs to obtain the approval.

5.7 Policies and Regulations of Radio Spectrum in the United Arab Emirates: A Comparative Perspective

This section provides an overview of the regulatory framework for radio equipment in the United Arab Emirates (UAE), highlighting the key differences and similarities with European standards. The aim is to shed light on how these regulations impact market access, technology adoption and deployment, and industry practices in the UAE, compared to Europe region.

The UAE has established itself as a regional hub for technological innovation and digital transformation. Central to this progress is the regulation of radio equipment, which ensures safety, compatibility, and efficient use of the radio spectrum. The UAE's regulatory framework is overseen by the Telecommunications and Digital Government Regulatory Authority (TDRA) [48], which mandates a *type approval* process. The UAE requires a Type Approval for all radio equipment before it can be marketed, sold or used in the UAE. This approach is outlined in the TDRA Policy for Telecommunications Equipment Type Approval Regime [49] and the TDRA Regulation for Telecommunications Equipment Type Approval Regime V3.0 [50]. These documents provide detailed guidelines and requirements for manufacturers and importers, ensuring that all telecommunications equipment complies with national standards, often based on the European Telecommunications Standards Institute (ETSI) standards [51], to guarantee safety, interoperability, and optimal use of the radio spectrum.

The approval process requires manufacturers or their authorized representatives to submit applications to TDRA, including detailed technical specifications and test reports. The equipment must comply with the TDRA's technical specifications. This mandatory approval process ensures that all equipment meets the UAE's technical standards and complies with national requirements. TDRA often adopts standards from ETSI for telecom devices and creates specific regulations for each device based on these benchmarks. While there are no specific dedicated laboratories in the UAE for all types of equipment, the country relies on internationally accredited laboratories that can test devices according to ISO/IEC 17,025 standards. The National Lab, under the TDRA's jurisdiction tests certain telecommunication devices, while TDRA issues certifications for devices based on their compliance with safety standards, electromagnetic compatibility (EMC), and radio frequency (RF) emission standards.

Importing radio equipment to the UAE requires customs clearance, adhering to TDRA's type approval process, in order to ensure compliance with national regulations. Importers must ensure that all radio equipment has received TDRA type approval before entering the UAE (such as mobile phones, private mobile radio devices (walkie talkies), etc.). Within this process, equipment must operate within allowed frequency bands and comply with TDRA's national requirements, including adherence to spectrum related regulations, one example is TDRA Regulations for Ultra-Wide Band (UWB) and Short-Range Devices (SRD) [52], as well as other specific technical conditions.

Following the initial approval, there are periodic re-assessments to ensure equipment continues to meet safety and emission standards, and to align with any regulatory updates or changes.

The UAE's spectrum management approach focuses on the efficient use of spectrum and supporting national priorities such as public safety and strategic communications. The National Frequency Plan [53] allocates frequency bands, which are controlled locally with considerations for sub-regional harmonization within the Gulf Cooperation Council (GCC) [54] and the Arab region through technical committees like the Arab Spectrum Management Group [55]. The TDRA publishes general regulations such as the "TDRA Regulations for Spectrum Allocation and Assignment" [56] and specific regulations for different types of radio services, including Private Mobile Radio [57], Fixed Radio Service [58], and Unmanned Aircraft Radio Systems [59]. Certain frequency bands and types of equipment which may cause potential interferences to existing services and uses require additional licensing. Equipment operating in frequency bands not compatible with the National Frequency Plan or with specific technical conditions may be restricted or require special permits. Restrictions are applied to equipment where the frequency bands have specific technical conditions specified in various spectrum-related regulations, such as maximum effective radiated power, magnetic field, channel access and occupation rules, and channel spacing or maximum occupied bandwidth.

While both the UAE and Europe aim to ensure the safe and efficient use of radio equipment, there are notable differences in their regulatory approaches. The European Radio Equipment Directive (RED) [2] emphasizes harmonization across member states to facilitate the free movement of goods within the EU's single market. The UAE's approach is more centralized, focusing on national priorities, safety, local compatibility, and cultural aspects. Devices approved under European standards, such as CE marking, might still require re-certification or modification to meet UAE regulations, particularly concerning frequency band allocation and power limits. In Europe, spectrum allocation often focuses on harmonization across multiple nations for seamless integration. The UAE's frequency allocations are more locally controlled, and some bands used in Europe may be restricted or have different technical conditions in the UAE. The UAE may impose restrictions on certain types of radio equipment, requiring additional licensing or adherence to specific technical conditions not present in the European system.

The UAE often follows ETSI and International Telecommunication Union (ITU) [60] standards, integrating them into national regulations with localized amendments as necessary. The country is working on initiatives to align its regulations with international standards through participation in global forums like the ITU. While interested in mutual recognition agreements to simplify market access, the UAE maintains distinct national requirements. The UAE focuses on harmonization within the GCC and Arab regions, participating in regional technical committees dedicated to spectrum management.

The UAE actively updates its regulations to address emerging technologies like 5G, Internet of Things (IoT), and smart city initiatives. TDRA ensures that radio equipment used in these technologies meets stringent standards, balancing technological innovation with safety and compliance. Embracing new technologies enhances the UAE's ICT digital infrastructure, aligning with national strategic objectives.

While the UAE's approach to radio equipment regulation presents significant opportunities, it also faces several challenges. Among these challenges are keeping pace with rapid technological advancements, ensuring that equipment meets security and privacy standards with the rise of connected devices, and balancing adherence to international standards while addressing specific local requirements and cultural aspects. Opportunities lie in closer alignment with international standards, which can simplify compliance for foreign manufacturers and encourage international trade. Supporting emerging technologies like 5G and IoT can position the UAE as a leader in technological innovation. Pursuing international mutual recognition agreements can streamline market access and reduce barriers for equipment manufacturers. Recommendations include continuing international collaboration by participating in global forums and technical committees to align regulations and share best practices, adapting regulations to accommodate new technologies while maintaining safety and compliance, and engaging with manufacturers and stakeholders to address challenges and facilitate market access.

In conclusion, the UAE's approach to market access for radio equipment is characterized by a centralized regulatory framework under TDRA, a mandatory type approval process, and proactive spectrum management. While there are differences from the European model, particularly in regulatory objectives and conformity assessment procedures, the UAE often aligns with international standards like those from ETSI and ITU, incorporating localized amendments to meet national priorities. Understanding these nuances is essential for manufacturers and stakeholders aiming to enter the UAE market, ensuring compliance and capitalizing on opportunities within this dynamic environment.

5.8 Conclusion

The European Radio Equipment Directive (RED) is the key regulatory framework governing access to the European Union Single Market for Radio Equipment. It is thus of utmost importance for manufacturers of such equipment to understand related essential requirements. If those are not met, market access may be lost. It is equally important for manufacturers of components of Radio Equipment to understand the regulatory framework. There is likely a key competitive advantage in case that components are designed such that related customers, such as manufacturers of final Radio Equipment products, are supported in their conformity assessment efforts.

Interestingly, the original focus of the Radio Equipment Directive has broadened since it was first published in 2014. Originally, the applicable essential requirements were mainly limited to the protection of health and safety, electromagnetic compatibility and the effective and efficient use of spectrum. Later, new Articles were invoked including the recent activation of Articles relating to Cybersecurity and Privacy requirements (Articles 3(3)(d/e/f)). Currently, the Article 3(3)(i) and Article 4 are under discussion with a corresponding Impact Assessment study being ongoing at the time when the present chapter was written. Those Articles are meant to provide protection against unauthorized Radio Software which may put a Radio Equipment into an uncompliant mode. Since almost any Radio Equipment involves Software affecting its radio characteristics, an understanding of corresponding requirements will be key to all manufacturers.

Finally, the future role of the Radio Equipment Directive with respect to the Cyber Resilience Act (CRA) [47], which was available as a Draft when the present chapter was written, needs to be clarified. As discussion in the section on *Interdependencies with other European Regulations* in the present chapter, there is an intention to move corresponding requirements and Harmonised Standards developed for Cybersecurity and Privacy related *essential requirements* under the CRA. Since the scope of the CRA is considerably broader compared to the Radio Equipment Directive, the exact implementation of this transfer needs to be closely followed by manufacturers.

Disclaimer: Note that the authors have assembled the information above to the best of their knowledge as available at the time when the present book was written. Any interpretation is the personal opinion of the authors. Also, the legislation framework can change and evolve any time. The only legally binding text is the one provided in the documents formally published in the European Union Official Journal.

Acknowledgements The authors would like to express their explicit thanks to Prof. Seungwon Choi of Hanyang University, Seoul and Vladimir Ivanov who have substantially contributed to the related standards work in ETSI TC RRS (until 2020).

References

1. Radio Equipment Directive, European Commission Website, available at https://single-market-economy.ec.europa.eu/sectors/electrical-and-electronic-engineering-industries-eei/radio-equipment-directive-red_en
2. DIRECTIVE 2014/53/EU OF THE EUROPEAN PARLIAMENT AND OF THE COUNCIL of 16 April 2014 on the harmonisation of the laws of the Member States relating to the making available on the market of radio equipment and repealing Directive 1999/5/EC
3. European Commission RED Guide, 2018, Version 19, available at https://ec.europa.eu/docsroom/documents/33162/attachments/1/translations/en/renditions/native
4. DIRECTIVE (EU) 2022/2380 OF THE EUROPEAN PARLIAMENT AND OF THE COUNCIL of 23 November 2022 amending Directive 2014/53/EU on the harmonisation of the laws of the Member States relating to the making available on the market of radio equipment (Text with EEA relevance)
5. COMMISSION DELEGATED REGULATION (EU) 2023/1717 of 27 June 2023 amending Directive 2014/53/EU of the European Parliament and of the Council as regards the technical specifications for the charging receptacle and charging communication protocol for all the categories or classes of radio equipment capable of being recharged by means of wired charging (Text with EEA relevance)

6. Standardisation request M/536, COMMISSION IMPLEMENTING DECISION of 4.8.2015 on a standardisation request to the European Committee for Electrotechnical Standardisation and to the European Telecommunications Standards Institute as regards radio equipment in support of Directive 2014/53/EU of the European Parliament and of the Council, C(2015)5376—Standardisation request M/536, available at https://ec.europa.eu/growth/tools-databases/enorm/mandate/536_en
7. ETSI EN 301 489-1 V2.2.3 (2019-11); ElectroMagnetic Compatibility (EMC) standard for radio equipment and services; Part 1: Common technical requirements; Harmonised Standard for ElectroMagnetic Compatibility, available at https://www.etsi.org/deliver/etsi_en/301400_301499/30148901/02.02.03_60/en_30148901v020203p.pdf
8. ETSI EG 203 367 V1.1.1 (2016-06), Guide to the application of harmonised standards covering articles 3.1b and 3.2 of the Directive 2014/53/EU (RED) to multi-radio and combined radio and non-radio equipment
9. ETSI EN 301 893 V2.1.1 (2017-05); 5 GHz RLAN; Harmonised Standard covering the essential requirements of article 3.2 of Directive 2014/53/EU
10. COMMISSION DELEGATED REGULATION (EU) 2022/30 of 29 October 2021 supplementing Directive 2014/53/EU of the European Parliament and of the Council with regard to the application of the essential requirements referred to in Article 3(3), points (d), (e) and (f), of that Directive (Text with EEA relevance), available at http://data.europa.eu/eli/reg_del/2022/30/2023-10-27
11. COMMISSION IMPLEMENTING DECISION on a standardisation request to the European Committee for Standardisation and the European Committee for Electrotechnical Standardisation as regards radio equipment in support of Directive 2014/53/EU of the European Parliament and of the Council and Commission Delegated Regulation (EU) 2022/30, available at https://ec.europa.eu/transparency/documents-register/detail?ref=C(2022)5637&lang=en
12. Standardisation request M/585 Amd 1: COMMISSION IMPLEMENTING DECISION of 23.8.2023 amending Implementing Decision C(2022) 5637 on a standardisation request to the European Committee for Standardisation and the European Committee for Electrotechnical Standardisation as regards radio equipment in support of Directive 2014/53/EU of the European Parliament and of the Council and Commission Delegated Regulation (EU) 2022/30, available at https://ec.europa.eu/growth/tools-databases/enorm/mandate/585Amd1_en
13. COMMISSION DELEGATED REGULATION (EU) 2019/320 of 12 December 2018 supplementing of Directive 2014/53/EU of the European Parliament and of the Council with regard to the application of the essential requirements referred to in Article 3(3)(g) of that Directive in order to ensure caller location in emergency communications from mobile devices
14. 2000/637/EC: Commission Decision of 22 September 2000 on the application of Article 3(3)(e) of Directive 1999/5/EC to radio equipment covered by the regional arrangement concerning the radiotelephone service on inland waterways
15. 2001/148/EC: Commission Decision of 21 February 2001 on the application of Article 3(3)(e) of Directive 1999/5/EC to avalanche beacons
16. 2013/638/EC: Commission Decision of 12 August 2013 on essential requirements relating to marine radio communication equipment which is intended to be used on non-SOLAS vessels and to participate in the Global Maritime Distress and Safety System (GMDSS)
17. 2005/53/EC: Commission Decision of 25 January 2005 on the application of Article 3(3)(e) of Directive 1999/5/EC of the European Parliament and of the Council to radio equipment intended to participate in the Automatic Identification System (AIS)
18. 2005/631/EC: Commission Decision of 29 August 2005 concerning essential requirements as referred to in Directive 1999/5/EC of the European Parliament and of the Council ensuring access of Cospas-Sarsat locator beacons to emergency services
19. COMMISSION IMPLEMENTING REGULATION (EU) 2017/1354 of 20 July 2017 specifying how to present the information provided for in Article 10(10) of Directive 2014/53/EU of the European Parliament and of the Council (Text with EEA relevance)
20. European Commission Expert Group Radio Equipment, information available at https://ec.europa.eu/transparency/expert-groups-register/screen/expert-groups/consult?lang=en&do=groupDetail.groupDetail&groupID=3587&Lang=BG

21. EN 18031-1:2024, Common security requirements for radio equipment - Part 1: Internet connected radio equipment, CEN/CENELEC
22. EN 18031-2:2024, Common security requirements for radio equipment - Part 2: radio equipment processing data, namely Internet connected radio equipment, childcare radio equipment, toys radio equipment and wearable radio equipment, CEN/CENELEC
23. EN 18031-3:2024, Common security requirements for radio equipment - Part 3: Internet connected radio equipment processing virtual money or monetary value, CEN/CENELEC
24. Regulation (EU) 2017/745 of the European Parliament and of the Council of 5 April 2017 on medical devices, amending Directive 2001/83/EC, Regulation (EC) No 178/2002 and Regulation (EC) No 1223/2009 and repealing Council Directives 90/385/EEC and 93/42/EEC, available at https://eur-lex.europa.eu/legal-content/EN/TXT/?uri=CELEX%3A32017R0745
25. Regulation (EU) 2017/746 of the European Parliament and of the Council of 5 April 2017 on in vitro diagnostic medical devices and repealing Directive 98/79/EC and Commission Decision 2010/227/EU, available at https://eur-lex.europa.eu/eli/reg/2017/746/oj
26. Regulation (EU) 2018/1139 of the European Parliament and of the Council of 4 July 2018 on common rules in the field of civil aviation and establishing a European Union Aviation Safety Agency, and amending Regulations (EC) No 2111/2005, (EC) No 1008/2008, (EU) No 996/2010, (EU) No 376/2014 and Directives 2014/30/EU and 2014/53/EU of the European Parliament and of the Council, and repealing Regulations (EC) No 552/2004 and (EC) No 216/2008 of the European Parliament and of the Council and Council Regulation (EEC) No 3922/91, available at https://eur-lex.europa.eu/eli/reg/2018/1139/oj
27. Regulation (EU) 2019/2144 of the European Parliament and of the Council of 27 November 2019 on type-approval requirements for motor vehicles and their trailers, and systems, components and separate technical units intended for such vehicles, as regards their general safety and the protection of vehicle occupants and vulnerable road users, amending Regulation (EU) 2018/858 of the European Parliament and of the Council and repealing Regulations (EC) No 78/2009, (EC) No 79/2009 and (EC) No 661/2009 of the European Parliament and of the Council and Commission Regulations (EC) No 631/2009, (EU) No 406/2010, (EU) No 672/2010, (EU) No 1003/2010, (EU) No 1005/2010, (EU) No 1008/2010, (EU) No 1009/2010, (EU) No 19/2011, (EU) No 109/2011, (EU) No 458/2011, (EU) No 65/2012, (EU) No 130/2012, (EU) No 347/2012, (EU) No 351/2012, (EU) No 1230/2012 and (EU) 2015/166, available at https://eur-lex.europa.eu/eli/reg/2019/2144/oj
28. Directive (EU) 2019/520 of the European Parliament and of the Council of 19 March 2019 on the interoperability of electronic road toll systems and facilitating cross-border exchange of information on the failure to pay road fees in the Union, available at https://eur-lex.europa.eu/eli/dir/2019/520/oj
29. ETSI EN 303 641 V1.1.2 (2020-03), Reconfigurable Radio Systems (RRS); Radio Equipment (RE) reconfiguration requirements
30. ETSI EN 303 648 V1.1.2 (2020-03), Reconfigurable Radio Systems (RRS); Radio Equipment (RE) reconfiguration architecture
31. ETSI EN 303 681-1 V1.1.2 (2020-03), Reconfigurable Radio Systems (RRS); Radio Equipment (RE) information models and protocols for generalized software reconfiguration architecture; Part 1: generalized Multi-radio Interface (gMURI)
32. ETSI EN 303 681-2 V1.1.2 (2020-03), Reconfigurable Radio Systems (RRS); Radio Equipment (RE) information models and protocols for generalized software reconfiguration architecture; Part 2: generalized Reconfigurable Radio Frequency Interface (gRRFI)
33. ETSI EN 303 681-3 V1.1.2 (2020-03), Reconfigurable Radio Systems (RRS); Radio Equipment (RE) information models and protocols for generalized software reconfiguration architecture; Part 3: generalized Unified Radio Application Interface (gURAI)
34. ETSI EN 303 681-4 V1.1.2 (2020-03), Reconfigurable Radio Systems (RRS); Radio Equipment (RE) information models and protocols for generalized software reconfiguration architecture; Part 4: generalized Radio Programming Interface (gRPI)
35. ETSI TR 103 585 V1.2.1 (2019-11), Reconfigurable Radio Systems (RRS); Radio Equipment (RE) reconfiguration use cases

36. ETSI EN 302 969 V1.3.1 (2018-05), Reconfigurable Radio Systems (RRS); Radio Reconfiguration related Requirements for Mobile Devices
37. ETSI EN 303 095 V1.3.1 (2018-05), Reconfigurable Radio Systems (RRS); Radio Reconfiguration related Architecture for Mobile Devices
38. ETSI EN 303 146-1 V1.3.1 (2018-06), Reconfigurable Radio Systems (RRS); Mobile Device Information Models and Protocols; Part 1: Multi-radio Interface (MURI)
39. ETSI EN 303 146-2 V1.2.1 (2016-06), Reconfigurable Radio Systems (RRS); Mobile Device (MD) information models and protocols; Part 2: Reconfigurable Radio Frequency Interface (RRFI)
40. ETSI EN 303 146-3 V1.3.1 (2018-06), Reconfigurable Radio Systems (RRS); Mobile Device (MD) information models and protocols; Part 3: Unified Radio Application Interface (URAI)
41. ETSI EN 303 146-4 V1.1.2 (2017-04), Reconfigurable Radio Systems (RRS); Mobile Device (MD) information models and protocols; Part 4: Radio Programming Interface (RPI)
42. ETSI TR 103 087 V1.2.1 (2017-11); Reconfigurable Radio Systems (RRS); Security related use cases and threats in Reconfigurable Radio Systems
43. ETSI TS 103 436 V1.2.1 (2018-02); Reconfigurable Radio Systems (RRS); Security requirements for reconfigurable radios
44. ETSI TS 103 850 V1.1.1 (2022-10); Reconfigurable Radio Systems (RRS); Definition of Radio Application Package
45. ETSI White Paper No. #57, Software Radio Reconfiguration: A modular Software Reconfiguration approach for radio equipment in general, 2nd Edition – May 2023, available at https://www.etsi.org/images/files/ETSIWhitePapers/ETSI-WP-57-A_modular_Software_Reconfiguration_approach_for_Radio_Equipment_in_general.pdf
46. General Approach for a Harmonised Standard in support of RED Article 3(3)(i), ETSI submission to EG RED, 11th February 2021, contacts ETSI TC RRS - Markus Mueck (Markus.Dominik.Mueck@intel.com), Paul Bender (Paul.Bender@BNetzA.de), Michael Gundlach (michael.gundlach@nokia.com), Marcello Pagnozzi (Marcello.Pagnozzi@etsi.org), ETSI document number RRS(21)053024, Expert Group RE document number EG RE (09)12
47. REGULATION OF THE EUROPEAN PARLIAMENT AND OF THE COUNCIL on horizontal cybersecurity requirements for products with digital elements and amending Regulations (EU) No 168/2013 and (EU) 2019/1020 and Directive (EU) 2020/1828 (Cyber Resilience Act), 25 September 2024, version adopted by the European Council on 10 October 2024, available at https://data.consilium.europa.eu/doc/document/PE-100-2023-INIT/en/pdf
48. Telecommunications and Digital Government Regulatory Authority (TDRA), available at www.tdra.gov.ae
49. TDRA Policy for Telecommunications Equipment Type Approval Regime; retrieved from https://tdra.gov.ae/-/media/About/Type-approval-PDF/Type-Approval-28102022/ITem-72-Type-approval-policy-21--EN.ashx?t=Policy%20%20Telecommunications%20Equipment%20Type%20Approval%20Regime
50. TDRA Regulation for Telecommunications Equipment Type Approval Regime V3.0; retrieved from https://tdra.gov.ae/-/media/About/Type-approval-PDF/Type-Approval-28102022/Item-73--Type-approval-regulation-21-EN.ashx?t=Regulation%20-%20Telecommunications%20Equipment%20Type%20Approval%20Regime%20V3.0
51. European Telecommunications Standards Institute (ETSI), available at https://www.etsi.org/
52. TDRA Regulations for Spectrum Allocation and Assignment; retrieved from https://tdra.gov.ae/-/media/About/regulations-and-ruling/EN/UWB-and-SRD-Regulations-50.ashx?t=TDRA%20Regulations%20for%20Ultra%20Wide%20Band%20and%20Short%20Range%20Devices%205.0
53. TDRA National Frequency Plan, available at https://tdra.gov.ae/-/media/About/regulations-and-ruling/EN/UAE-National-Frequency-Plan-new-2020-En-pdf.ashx?t=UAE%20National%20Frequency%20Plan%202020
54. Gulf Cooperation Council, available at https://www.gcc-sg.org/

55. Arab Spectrum Management Group (ASMG, 2024); retrieved from ASMG: https://www.google.ae/url?sa=t&rct=j&q=&esrc=s&source=web&cd=&cad=rja&uact=8&ved=2ahUKEwjD55Se5--IAxXlBdsEHXzfA2wQFnoECBUQAQ&url=https%3A%2F%2Fasmg32.org%2F&usg=AOvVaw0OMSbO82NJysLrkRr5ut7U&opi=89978449
56. TDRA Regulations for Spectrum Allocation and Assignment; retrieved from https://tdra.gov.ae/-/media/About/regulations-and-ruling/EN/Spectrum-Allocation-and-Assignment-Regulations-V2-0-pdf.ashx?t=Spectrum%20Allocation%20and%20Assignment%20Regulations%202.0
57. TDRA Regulations for Private Mobile Radio (PMR); retrieved from https://tdra.gov.ae/-/media/About/regulations-and-ruling/Updated-regulations-and-ruling/Private-Mobile-Radio-PMR-Regs-v30-English.ashx?t=Private%20Mobile%20Radio%20(PMR)%20Regulations
58. TDRA Regulations for Fixed Radio Service; retrieved from https://tdra.gov.ae/-/media/About/regulations-and-ruling/Updated-regulations-and-ruling/Private-Mobile-Radio-PMR-Regs-v30-English.ashx?t=Private%20Mobile%20Radio%20(PMR)%20Regulations
59. TDRA Regulations for Unmanned Aircraft Radio Systems; retrieved from https://tdra.gov.ae/-/media/About/regulations-and-ruling/EN/TDRA-Regulations-on-Unmanned-Aircraft-Radio-Systems-v-20--28-Dec-2022.ashx?t=TDRA%20Regulations%20for%20Unmanned%20Aircraft%20Radio%20Systems%202.0
60. International Telecommunications Union (ITU), available at http://www.itu.int
61. Guide to the Radio Equipment Directive 2014/53/EU, 18th December 2018, available at https://ec.europa.eu/docsroom/documents/33162
62. Commission Delegated Regulation (EU) 2022/30 of 29 October 2021 supplementing Directive 2014/53/EU of the European Parliament and of the Council with regard to the application of the essential requirements referred to in Article 3(3), points (d), (e) and (f), of that Directive (Text with EEA relevance)

Chapter 6
Introduction to the European Intelligent Transport Systems Regulation

Markus Mueck and Christophe Gaie

Abstract Intelligent Transport Systems (ITS) are proven to lead to a cleaner, safer and more efficient transport system. Since the implementation of the European ITS Directive 2010/40/EU on 7th July 2010, the road management efficiency has been increased and road accident fatalities in Europe have indeed dropped significantly. Implemented priorities include traffic and travel information, the eCall emergency system and intelligent truck parking, followed by real-time traffic information and multimodal travel information. Further efficiency gains and a significant drop in road accident fatalities are expected through new measures introduced by Directive (EU) 2023/2661 which was adopted on the 22nd November 2023 and is amending the original Directive 2010/40/EU. The goal is to facilitate the adoption of emerging road mobility options, mobility apps, and connected and automated mobility enabled by Cooperative Intelligent Transport Systems (C-ITS). To achieve this, the proposal suggests making certain essential road, travel, and traffic data accessible in digital format, such as speed limits, traffic circulation plans, and roadworks. This will encourage the faster deployment of innovative, intelligent services. In this context, Vehicle-to-Everything (V2X) communication technology is expected to act as a key enable of such connected services. V2X technology has been developed starting over 15 years ago and has achieved considerable maturity. The present chapter gives an overview on the status of ITS and C-ITS regulation in Europe and provides an overview on the underlying communication technology.

Keywords Regulation · Vehicular communication · Cyber security · Privacy

---

M. Mueck (✉)
Intel Deutschland GmbH, Munich, Germany
e-mail: Markus.Dominik.Mueck@intel.com

C. Gaie
Head of the Engineering and Digital Innovation Division, Prime Minister Services, Paris, France
e-mail: christophe.gaie@gmail.com

M. Mueck and C. Gaie (eds.), *European Digital Regulations*, Intelligent Systems Reference Library 265, https://doi.org/10.1007/978-3-031-80809-8_6

6.1 Introduction

6.1.1 V2X in the Context of EU Regulation

Intelligent Transport Systems (ITS) are actively supported through related regulation efforts around the world. Europe has been on the forefront of introducing related features [1] through the European ITS Directive 2010/40/EU [2] adopted on 7th July 2010 including traffic and travel information, the eCall emergency system and intelligent truck parking, followed by real-time traffic information and multimodal travel information. As it will be further detailed in the following sections of the present chapter, a number of delegated Acts under Directive 2010/40/EU [2] were subsequently implemented, for example targeting a harmonised provision for an interoperable EU-wide eCall, the provision of road safety-related minimum universal traffic information free of charge to users, etc. Since then, road accident fatalities have dropped significantly across Europe [3].

In the mean-time, the technology has been evolving and new enablers for Cooperative Intelligent Transport Systems (C-ITS) are becoming available. Especially, Vehicle-to-Everything (V2X) communication technology is such a key enabler for C-ITS. It has been becoming mature and is enabling a vast series of Use Cases to increase overall traffic management efficient and road safety, as for example defined by the European Telecommunications Standards Institute (ETSI)—including Release 2 of the *Basic Set of Applications* [4] -, the 5G Automotive Association—including *C-V2X Use Cases and Service Level Requirements*, Volume III [5]—and others. Market adoption of V2X technology, however, was slower than originally expected. In the view of the authors of the present chapter, one reason relates to the fact that two different and mutually incompatible access layer technologies are competing against each other for usage in the 5.9 GHz band [6, 7]:

- ITS-G5 [8] whose access layer was originally developed by the Institute of Electrical and Electronics Engineers (IEEE) as a derivate of WiFi technology and published in 2010: the IEEE Std 802.11p-2010 [9];
- Cellular Vehicle-to-Everything (C-V2X) [10], also named LTE-V2X for 4th Generation cellular technology and NR-V2X (New Radio-V2X) for 5th Generation cellular technology, which is defined by the 3rd Generation Partnership Project (3GPP) and tightly integrated into the Cellular Communications ecosystem.

Note that the term ITS-G5 is used in Europe, while other regions apply different terminology. For example, in the United States the terms Wireless Access in Vehicular Environments (WAVE) and Dedicated Short Range Communications (DSRC) are used instead. While the access layers are different for both technologies, ITS-G5 and C-V2X, upper layers are identical—except for some technology specific features such as Decentralized Congestion Control (DCC) [11] required for ITS-G5 while C-V2X inherently avoids congestion issues and does thus not require DCC - and defined by ETSI for the application in Europe. Other flavours of the upper layer solutions are defined by other regions individually.

In Europe, the European Commission, the European Parliament and the European Council had been working on a Delegated Act [12] supplementing Directive 2010/40/EU with the objective to mandate the deployment of ITS-G5 technology across Europe. The Council of the European Union objected to the proposal for Delegated Regulation on Cooperative Intelligent Transport Systems (C-ITS) on July 8, 2019, at the final stage of the legislative process [13]. This objection was due to differing views on the best technology choice.

In the mean-time, the United States have changed the regulation moving from an original adoption of DSRC/WAVE to C-V2X. On July 17, 2024, the United States Federal Communications Commission (FCC) announced to schedule a vote *to finalize the Transition of Intelligent Transportation Systems to C-V2X Technology in the 5.9 GHz Band* [14]. China exclusively adopts C-V2X [15]. Europe currently adopts a technology neutral approach [16], allowing for the usage of both technologies, ITS-G5 and C-V2X.

In November 2023, the European Union updated its Intelligent Transport Systems (ITS) Directive (2010/40/EU) with Directive (EU) 2023/2661 [2]. This update aims to support the growth of new mobility options, apps, and connected and automated vehicles. A key goal is to accelerate the deployment of innovative, intelligent services. To achieve this, the directive proposes making certain critical road, travel, and traffic data available in digital format. This includes information like speed limits, traffic flow plans, and road construction updates [1].

Furthermore, Directive (EU) 2023/2661 expands the scope of ITS to include Cooperative Intelligent Transport Systems (C-ITS). C-ITS enables vehicles to communicate and cooperate with each other, sharing secure and reliable information without prior knowledge or discrimination [17]. This expansion is expected to foster more advanced and efficient transportation solutions.

The following sections will provide an overview of the ITS Directive 2010/40/EU [2] including its amendment Directive (EU) 2023/2661 [17] as well as related working programmes and Delegated Acts. Also, some key aspects of V2X technology will be discussed including security mechanisms and spectrum allocation.

6.1.2 A Deeper Dive into V2X Technology

V2X, or Vehicle-to-Everything, is a revolutionary communication technology that enables vehicles to exchange data with other vehicles, infrastructure, and pedestrians as illustrated in Fig. 6.1 inspired from [18]. This interconnectivity is a key aspect to develop safer, more efficient, and sustainable transports.

At the heart of V2X lies the Onboard Unit (OBU), an in-vehicle communication unit, attached with the vehicular control system and ensuring the communication interface with other entities [19]. This device transmits and receives V2X messages with various information such as vehicle speed, location, and braking status. By

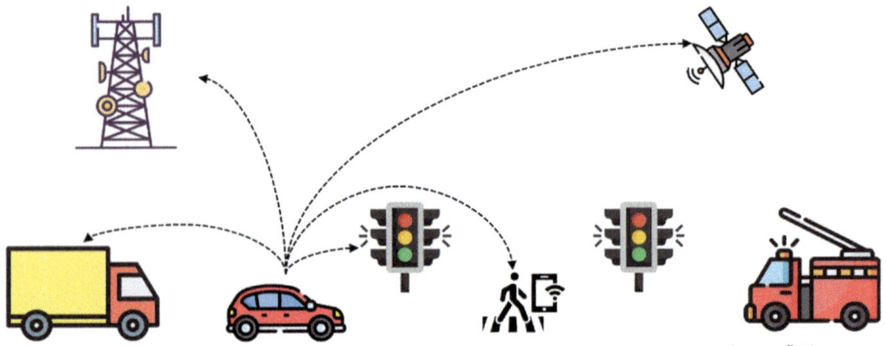

Fig. 6.1 Principle of vehicle-to-everything

sharing this data, vehicles can anticipate each other's movements, reducing the risk of collisions and improving traffic flow.

V2X also facilitates communication between vehicles and infrastructure. Roadside units (RSUs) can broadcast information about traffic conditions, hazards, and road closures to nearby vehicles. This enables drivers to make informed decisions and avoid potential dangers to achieve zero road fatalities by 2050 [20]. Conversely, vehicles can send information to RSUs, such as reporting accidents or road conditions that may affect other drivers.

Beyond vehicle-to-vehicle and vehicle-to-infrastructure communication, V2X also has the potential to improve pedestrian safety. By equipping pedestrians with devices that can communicate with nearby vehicles, it becomes possible to provide alerts and warnings to pedestrians who may be in danger. This could significantly reduce the number of pedestrian accidents using Wifi communications such as detailed by Liu et al. [21].

To enable V2X communication, a set of standards and protocols has been developed. IEEE 802.11p, also known as ITS-G5, is one of the primary standards for V2X. It is based on the Wi-Fi standard and provides a reliable and efficient way for vehicles to communicate with each other. Other standardizations were defined by 3GPP based on LTE in Release 14 (LTE-V2X) and 5G in Release 15 (5G-V2X). In either case, C-V2X is the generic terminology that refers to the V2X technology using cellular technology, regardless of the specific generation [22].

V2X technology offers substantial improvements in road safety and traffic efficiency; however, prioritizing security is essential to mitigate potential vulnerabilities. Protecting V2X communications from cyberattacks is crucial for maintaining the integrity and reliability of intelligent transportation systems (ITS) applications. To safeguard the privacy of vehicle data and prevent interference from malicious actors, the implementation of robust security measures such as encryption, authentication, and intrusion detection is necessary [23].

The widespread adoption of vehicle-to-everything (V2X) technology faces several challenges, including spectrum allocation, costs, and security concerns [24].

However, as V2X evolves toward the Internet of Vehicles (IoV) to meet the growing demands of advanced vehicular applications, these challenges are being progressively addressed through technological advancements and policy developments.

6.2 European ITS Regulation

As discussed above, the European ITS Directive 2010/40/EU [2] adopted on 7th July 2010 and amended by Directive (EU) 2023/2661 [17] on 22nd November 2023. A summary of the key objectives and related Delegated Acts will be given below.

6.2.1 The European ITS Directive 2010/40/EU

The European ITS Directive 2010/40/EU establishes a framework to support the coordinated and coherent deployment and use of Intelligent Transport Systems (ITS) across the EU, especially between Member States [2]. It outlines the general conditions needed for this purpose. In this context, the following four priority areas are introduced as illustrated in Fig. 6.2.

The four priority areas of the European ITS Directive 2010/40/EU [2] are summarized below:

1. **Priority area I: Optimal use of road, traffic and travel data**
   - Specifications and standards must be developed to ensure the optimal use of road, traffic and travel data;

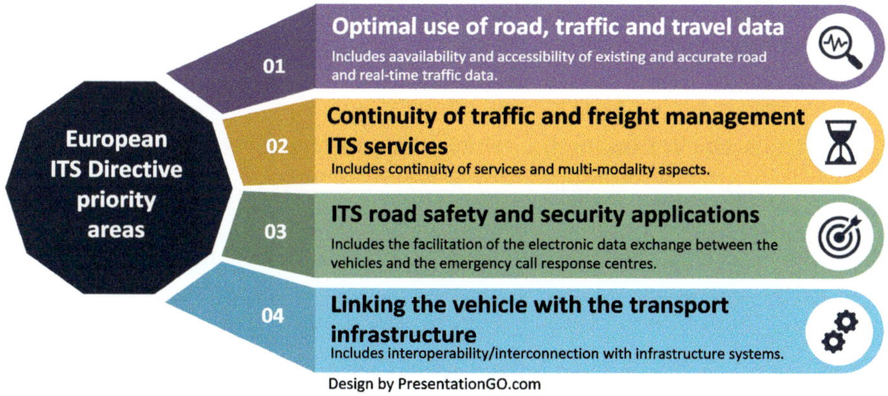

Fig. 6.2 The four priority areas of the European ITS directive 2010/40/EU [2]

- Availability and accessibility of existing and accurate road and real-time traffic data need to be ensured to be used for multimodal travel information to ITS service providers;
- Facilitating electronic data exchange between relevant public authorities, stakeholders, and ITS service providers across borders is essential;
- Updating available road and traffic data at a timely frequency will facilitate multimodal travel information.

2. **Priority area II: Continuity of traffic and freight management ITS services**

   - Defining the necessary measures to develop an EU ITS Framework Architecture is crucial. This architecture should specifically address ITS-related interoperability, continuity of services, and multi-modality aspects, such as multimodal interoperable ticketing;
   - To ensure the continuity of ITS services, particularly for cross-border passenger transport, minimum necessary requirements must be established for managing passenger transport across various modes of transportation;
   - For the seamless management of freight along transport corridors and across different modes of transport, minimum necessary requirements for ITS service continuity must be defined;
   - To facilitate freight transport logistics (eFreight), necessary measures should be implemented to realize ITS applications, such as tracking and tracing freight throughout its journey and across different modes of transportation;
   - To guarantee interoperability and compatibility between urban and European ITS architectures, essential interfaces must be defined.

3. **Priority area III: ITS road safety and security applications**

   - To ensure a harmonized, interoperable eCall system across the EU, necessary measures must be defined;
   - This includes specifying the required in-vehicle ITS data to be exchanged, ensuring the availability of necessary equipment in emergency call response centers, and facilitating electronic data exchange between vehicles and these centers.

4. **Priority area IV: Linking the vehicle with the transport infrastructure**

   - To create an open in-vehicle platform, it is essential to define the necessary measures for integrating various ITS applications.
   - To advance the development and implementation of cooperative systems (vehicle-to-vehicle, vehicle-to-infrastructure, and infrastructure-to-infrastructure), necessary measures must be implemented.

Furthermore, Article 3 of the European ITS Directive 2010/40/EU [2] derives the *priority actions* depicted in Fig. 6.3:

Fig. 6.3 Priority Actions introduced by Article 3 of the European ITS directive 2010/40/EU [2]

6.2.2 Directive (EU) 2023/2661 Amending the European ITS Directive 2010/40/EU

The European ITS Directive 2010/40/EU [2] was adopted on 7th July 2010 and amended by Directive (EU) 2023/2661 [17] on 22nd November 2023. Directive (EU) 2023/2661 [17] provides for the availability of data and the deployment of ITS services within the new priority areas illustrated in Fig. 6.4:

The four priority areas of Directive (EU) 2023/2661 [17] amending the European ITS Directive 2010/40/EU [2] are summarized below:

1. **Priority area I: information and mobility ITS services**
   Here's an optimized version:
   - Specifications should be developed for EU-wide multimodal digital mobility services, including EU-wide multimodal travel information services;
   - Specifications have to be set for EU-wide road traffic information and navigation services, including real-time traffic information services;
   - Specifications are necessary for both EU-wide multimodal digital mobility services and road traffic information and navigation services.
2. **Priority area II: Travel, transport and traffic management ITS services**
   - Specifications are required for enhanced traffic and incident management services;

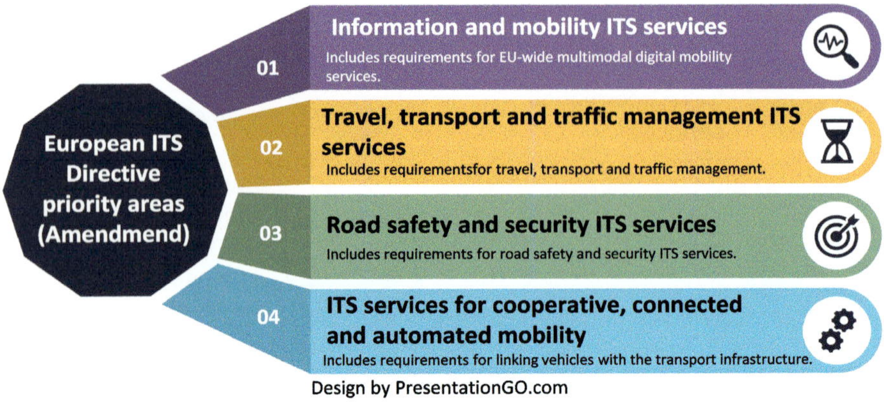

Fig. 6.4 The four priority areas introduced by directive (EU) 2023/2661 [17] amending the European ITS directive 2010/40/EU [2]

- Mobility management services necessitate the development of specifications;
- An EU-wide ITS architecture framework should be established;
- ITS applications and freight transport logistics need to be clearly defined.

3. **Priority area III: Road safety and security ITS services**

   - Specifications are required for an interoperable EU-wide eCall system;
   - Information and reservation services for safe and secure parking spaces for trucks and commercial vehicles need specifications;
   - Road safety-related minimum universal traffic information specifications are necessary;
   - Other actions require specifications, including:
     - Supporting the safety of road users through on-board Human–Machine Interfaces and nomadic device usage, including cell phones, to aid driving tasks and transport operations, as well as in-vehicle communication security.
     - Implementing necessary measures to enhance the safety and comfort of vulnerable road users.
     - Integrating advanced driver support information systems into vehicles and road infrastructure.
     - Facilitating information exchange between security ITS application service providers.

4. **Priority area IV: ITS services for cooperative, connected and automated mobility**

   - Necessary measures should be defined to advance the development and implementation of cooperative intelligent transport systems (C-ITS), focusing on supporting Connected, Cooperative & Automated Mobility (CCAM).

- Specifications are required to enable services such as:
  - C-ITS information and warning services based on status data to enhance transport user awareness of upcoming traffic situations.
  - C-ITS information and warning services based on observations to further increase the awareness of transport users, including non-connected users, of upcoming traffic situations.
  - C-ITS services based on intentions to facilitate automated driving in complex traffic scenarios.
  - C-ITS infrastructure services to support automated driving.
- Specifications should be developed for the EU C-ITS security credential management system.

6.2.3 Work Programme

The European Commission has published three Commission Decisions related to a Working Programme on the implementation of Directive 2010/40/EU:

1. **On February 15, 2011, the European Commission Adopted a Working Programme to Implement Directive 2010/40/EU** [25]

The working program includes a detailed timing and planning proposal for the implementation of the solutions illustrated in Fig. 6.5:

The six activities indicated above are expected to be performed during the period 2011–2015 as regards the preparation and development of specifications related to the six Priority Actions as defined in Article 3 of the ITS Directive. For each of the items, the following is defined:

- Time frame for analysis and preparation;
- Time frame for impact assessment;
- Time frame for drafting of specifications;
- Time frame for availability of final draft and inter-services consultation;

Fig. 6.5 Six activities to be performed during the period 2011–2015 [25] in support of Directive 2010/40/EU [2]

- Time frame for adoption.

2. **On December 11, 2018, the European Commission Adopted a Working Programme in Relation to the Actions Under Article 6(3) of Directive 2010/40/EU** [26]

The working program includes a detailed timing and planning proposal for the implementation of the solutions illustrated in Fig. 6.6:

3. **On December 13, 2022, the European Commission Updated the Working Programme in Relation to the Actions Under Article 6(3) of Directive 2010/40/EU for the Period 2022–2027** [27]

The working program includes a detailed timing and planning proposal for the implementation of the solutions illustrated in Fig. 6.7:

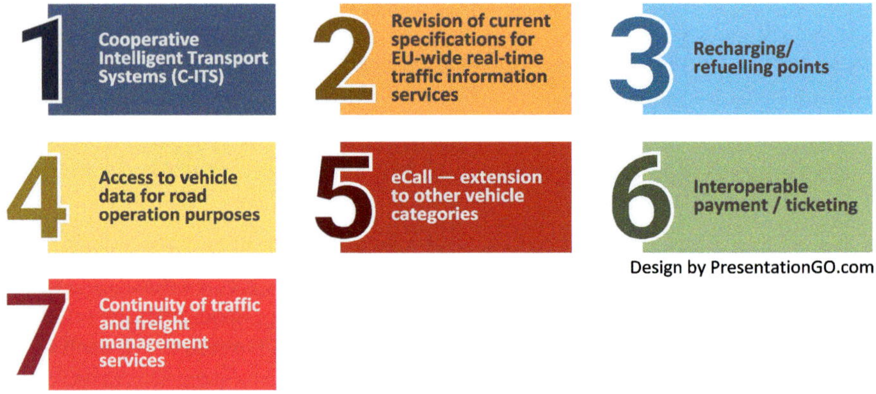

Fig. 6.6 Seven activities to be performed during the period 2018–2022 [26] in support of Directive 2010/40/EU [2]

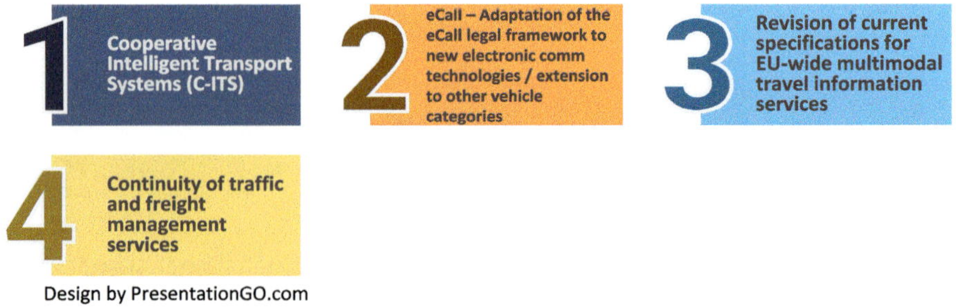

Fig. 6.7 Four activities to be performed during the period 2022–2027 [27] in support of Directive 2010/40/EU [2]

Currently, a Working Programme is under development for the time period 2024–2028. In a related Workshop presentation [28], the European Commission outlines the following planned initiatives:

1. **Priority area I—Information and mobility ITS services**
   - Envisaged initiative: Development of multimodal access nodes identifiers.
2. **Priority area II—Travel, transport and traffic management ITS services**
   - Envisaged initiative: Continuity of traffic and freight management services—Enhanced traffic management.
3. **Priority area III—Road safety and security ITS services**
   - Envisaged initiative: Extension of eCall to other categories of vehicles;
   - Envisaged initiative: Revision of the SRTI (Safety-Related Traffic Information) Delegated Regulation;
   - Envisaged initiative: Revision of the SSTP (Safe and Secure Truck Parking) Delegated Regulation.
4. **Priority area IV—ITS services for cooperative, connected and automated mobility**
   - Envisaged initiative: Specifications for the EU C-ITS security credential management system;
   - Envisaged initiative: Specifications for harmonised C-ITS services;
   - Envisaged initiative: Specifications to foster the development and implementation of C-ITS.

6.2.4 Delegated Acts

The following Delegated Acts under Directive 2010/40/EU [2] have been published to support its implementation as depicted in Fig. 6.8.

6.3 Vehicle-To-Everything (V2X) Communication

As discussed earlier in the present chapter, V2X systems offer a vast variety of services including improvement of road safety and traffic efficiency, protection of vulnerably road users, advanced infotainment, etc. Moreover, V2X may relate to Vehicle-to-Vehicle (V2V) communication, Vehicle-to-Pedestrian (V2P), Vehicle-to-Infrastructure (V2I), Vehicle-to-Network (V2N), etc. The corresponding V2X system thus comprises a multitude of entities interacting, including On-Board Units (OBUs) within the vehicle, Road Side Units (RSUs) at the road side, possibly applications running on Edge nodes or remote Cloud Data Centers, etc.

Fig. 6.8 Delegated acts under directive 2010/40/EU [2]

The Access Layer designs applicable for usage in Europe are summarized for ITS-G5 in [8] and for Cellular Vehicle-to-Everything (C-V2X) in [10]. In the following, we are discussing some key components of V2X: The overall protocol stack of ITS-G5 and C-V2X respectively, Cybersecurity approaches for V2X and spectrum allocated in Europe in the 5.9 GHz band.

6.3.1 The European Protocol Stack of ITS-G5 and C-V2X

The protocol stack of ITS-G5 builds on IEEE 802.11p technology and is summarized by [29] in Fig. 6.9 as follows:

The protocol stack of C-V2X builds on 3GPP technology and is summarized by [30] and illustrated in Fig. 6.10:

6.3.2 Cybersecurity for V2X

The complex and multi-faceted set-up of V2X technology offers a variety of attack vectors, including attacks on a hardware level, attacks targeting information and data to be exchanged, attacks related to the over-the-air communication system, etc. It is thus obvious that a complex Ende-to-End security approach is required covering such attack scenarios.

In the present chapter, the authors choose to focus on two aspects of security: Application-Layer and Access-Layer security. Those are of specific interest, because a fully standardised and industry-agreed solution needs to be developed such that equipment originating from distinct manufacturers remain interoperable. Other parts

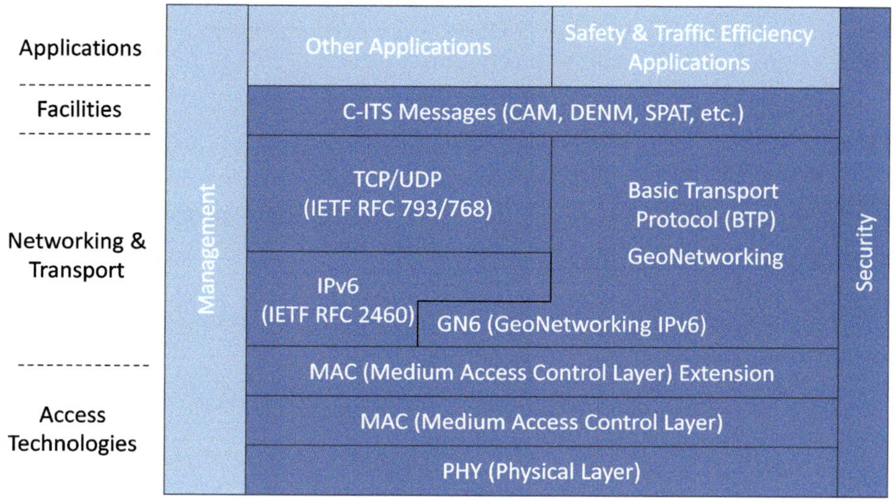

Fig. 6.9 The protocol stack of ITS-G5 [29]

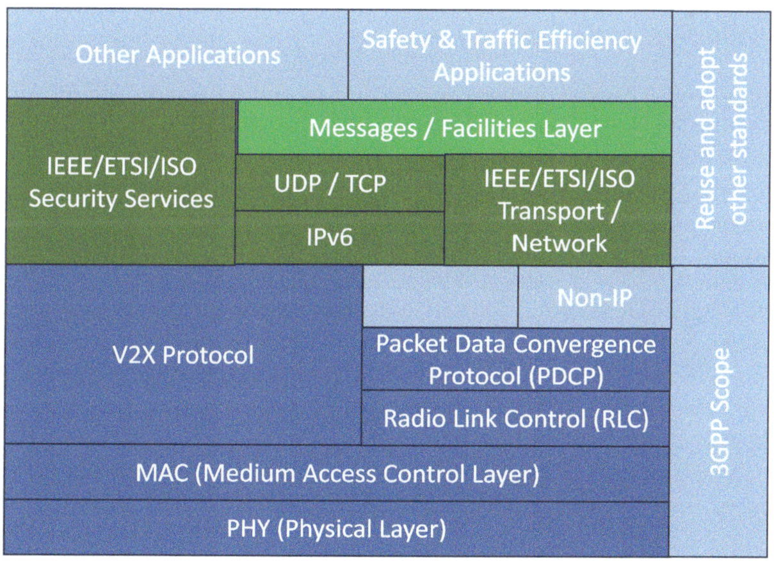

Fig. 6.10 The protocol stack of C-V2X [30]

of the chain may rely on design security principles, typically building on Functional Safety principles as defined by ISO 26262 [31].

Concerning Access-Layer security, i.e. covering the security related to the over-the-air communication, we need to differentiate solutions for the three prevailing technologies: WAVE/ITS-G5 (IEEE 802.11p based solution for US/Europe), 3GPP

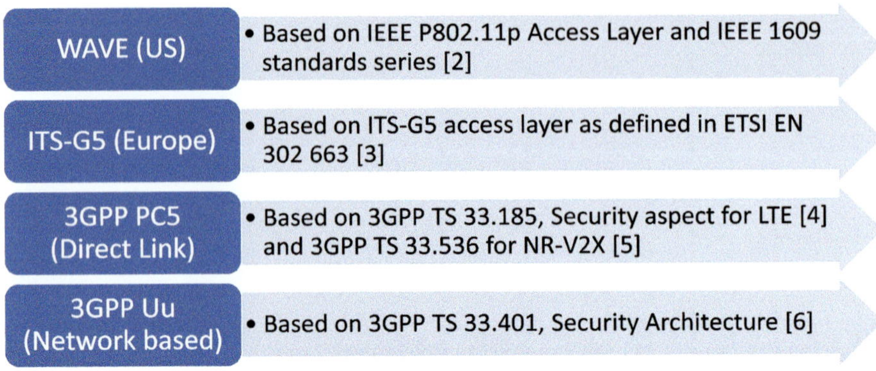

Fig. 6.11 Access layer security for V2X

PC5 (cellular direct communication) and 3GPP Uu (cellular network-based communication) as illustrated in Fig. 6.11.

The V2X security system architecture is building on the public key infrastructure (PKI) approach defined in IEEE 1609.2 across the United States, Europe and China with specific regional adaptations.

In the United States, a Security Credential Management System (SCMS) is defined by IEEE 1609.2.1 which introduces a security certificate management system. This is complemented by a number of SAE standards, for example SAE J2735 on a Data Dictionary, SAE J2945.x Interface standards and others as explained in [32]. The SCSM is discussed in detail by the US Department of Transportation in document [33]. A summary illustration is given in Fig. 6.12.

In Europe, ETSI TS 102 941 [34] defines the C-ITS Security Credential Management System (CCMS). It is a security certificate management system with considerable differences over SCMS employed in the United States. CCMS handles different trust models, allows one or more root CAs to exist, and realizes multiple certificate management modes such as single root CA, cross authentication, bridging CA, and certificate trust list [35]. The illustration of Fig. 6.13 summarizes the European framework.

The upper Security Credential Management schemes rely on a number of distinct certificates which are summarized in Fig. 6.14.

6.3.3 Spectrum Allocation for V2X in Europe

As summarized in the 5GAA White Paper [36], European administrations have designated the 5.9 band, and more specifically the bands 5855–5875 MHz and 5875–5925 MHz, for use by road Intelligent Transport Systems (ITS) as specified in ECC Recommendation (08)01 [37] and ECC Decision (08)01 [38]. These are complemented by Decision 2020/1426 [7] of the European Commission. As is common

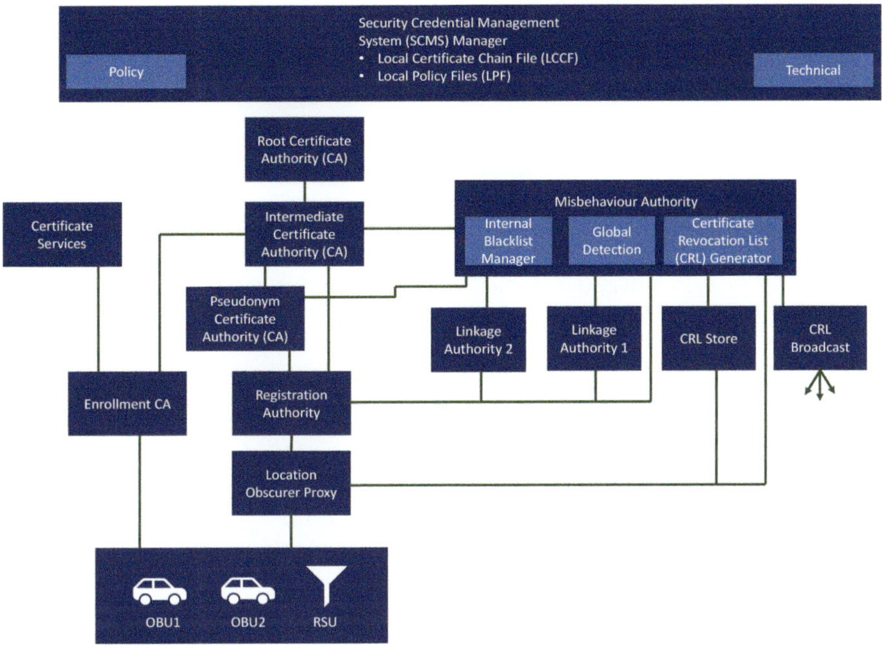

Fig. 6.12 The United States security credential management system (SCMS) [33]

practice in Europe, the spectrum is designated on a technology neutral basis. The allocation is illustrated in Fig. 6.15.

Furthermore, tolling (electronic fee collection) systems are operating in the 5 795 MHz to 5 815 MHz band based on the CEN DSRC standard [39, 40]. Interestingly, there is an ongoing debate on possible interference from V2X communication systems operating in the 5.9 GHz band to CEN DSRC systems although there is a considerable separation between the bands. In the view of the authors, well designed CEN DSRC systems should not encounter any issues. It is likely that early deployments of CEN DSRC systems in some European countries have neglected the possibility that the 5.9 GHz band may be used in the future and thus inadequate product design assumptions have been taken, violating basic principles of efficient and effective use of spectrum.

While the current 5.9 GHz channel allocation is relying on a 10 MHz channelization, there is currently a discussion ongoing whether to allow 20 MHz channels which will lead to a more efficient operation of C-ITS systems. Indeed, the review of ECC Decision (08)01 [37], which deals with the harmonised use of Safety-Related Intelligent Transport Systems (ITS) in the 5875–5935 MHz frequency band, was initiated [41].

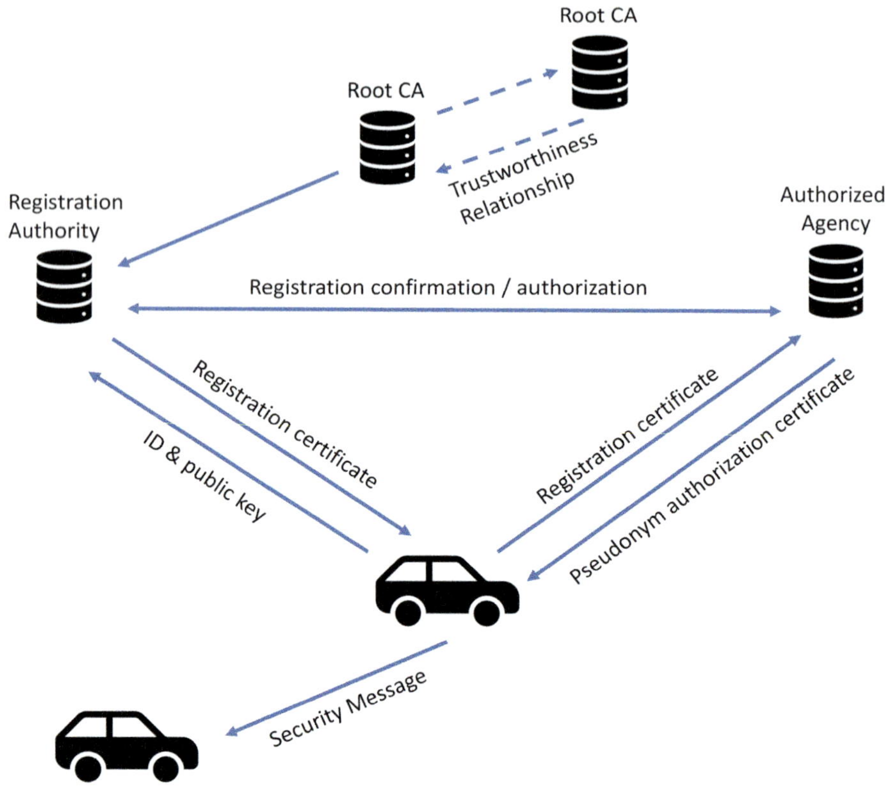

Fig. 6.13 European security credential management system (CCMS) architecture (European Commission 2018)

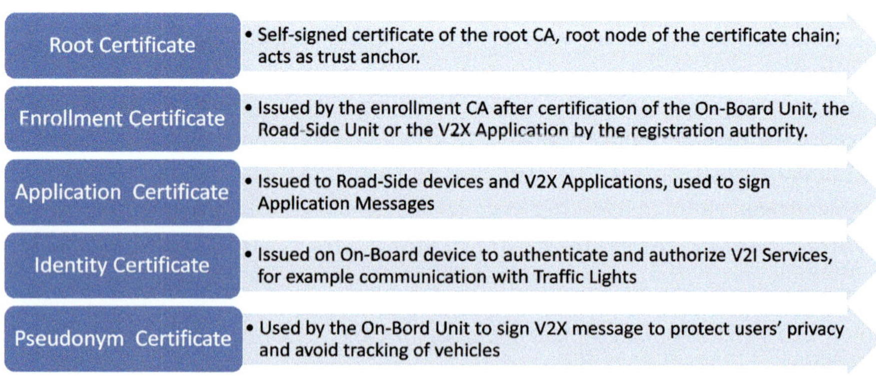

Fig. 6.14 Summary of certificate types [35]

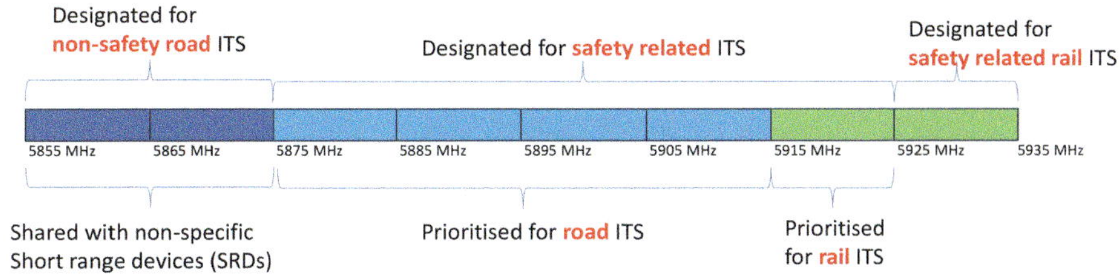

Fig. 6.15 V2X 5.9 GHz band allocation in Europe

6.4 Conclusion

Intelligent Transport Systems (ITS) contribute substantially to more efficient and safer transport systems. Recently, the original ITS principles are evolving towards Cooperative Intelligent Transport Systems (C-ITS). In this context, V2X communication systems, based on ITS-G5 or C-V2X technology, are a key component to C-ITS introducing the possibility of data exchange between vehicles, vehicles and pedestrians, vehicles and infrastructure, etc. Besides significant contributions in terms of sustainability to Europe's Green Deal, it is expected that road traffic fatalities will be further reduced dramatically. Hereby, modern C-ITS solutions are strongly relying on communication technology and need to be considered from the perspective of data management, data privacy, cybersecurity and similar.

Disclaimer: *Note that the authors have assembled the information above to the best of their knowledge as available at the time when the present book was written. Any interpretation is the personal opinion of the authors. Also, the legislation framework can change and evolve any time. The only legally binding text is the one provided in the documents formally published in the European Union Official Journal.*

References

1. ITS Directive and Action Plan, European Commission, available at https://transport.ec.europa.eu/transport-themes/smart-mobility/road/its-directive-and-action-plan_en
2. Directive 2010/40/EU of the European Parliament and of the Council of 7 July 2010 on the framework for the deployment of Intelligent Transport Systems in the field of road transport and for interfaces with other modes of transport Text with EEA relevance, available at https://eur-lex.europa.eu/legal-content/EN/TXT/?uri=CELEX%3A32010L0040
3. Road accident fatalities, available at https://ec.europa.eu/eurostat/statistics-explained/index.php?title=Road_safety_statistics_in_the_EU&oldid=630784
4. ETSI TR 102 638 V2.1.1 (2024–04): Intelligent Transport Systems (ITS); Vehicular Communications; Basic Set of Applications; Release 2, available at https://www.etsi.org/deliver/etsi_tr/102600_102699/102638/02.01.01_60/tr_102638v020101p.pdf
5. C-V2X Use Cases and Service Level Requirements—Volume III, 5G Automotive Association, January 2023, available at https://5gaa.org/c-v2x-use-cases-and-service-level-requirements-volume-iii/

6. 2008/671/EC: Commission Decision of 5 August 2008 on the harmonised use of radio spectrum in the 5875 - 5905 MHz frequency band for safety-related applications of Intelligent Transport Systems (ITS) (notified under document number C(2008) 4145) (Text with EEA relevance), available at https://eur-lex.europa.eu/eli/dec/2008/671/oj
7. Commission Implementing Decision (EU) 2020/1426 of 7 October 2020 on the harmonised use of radio spectrum in the 5 875–5 935 MHz frequency band for safety-related applications of intelligent transport systems (ITS) and repealing Decision 2008/671/EC (notified under document C(2020) 6773) (Text with EEA relevance), available at https://eur-lex.europa.eu/eli/dec_impl/2020/1426/oj
8. ETSI EN 302 663 V1.3.1 (2020–01): Intelligent Transport Systems (ITS); ITS-G5 Access layer specification for Intelligent Transport Systems operating in the 5 GHz frequency band, available at https://www.etsi.org/deliver/etsi_en/302600_302699/302663/01.03.01_60/en_302663v010301p.pdf
9. IEEE 802.11p-2010; IEEE Standard for Information technology-- Local and metropolitan area networks-- Specific requirements-- Part 11: Wireless LAN Medium Access Control (MAC) and Physical Layer (PHY) Specifications Amendment 6: Wireless Access in Vehicular Environments, available at https://standards.ieee.org/ieee/802.11p/3953
10. ETSI EN 303 798 V2.1.1 (2024–08): Intelligent Transport Systems (ITS); LTE-V2X and NR-V2X Access layer specification for Intelligent Transport Systems operating in the 5 GHz frequency band; Release 2, available at https://www.etsi.org/deliver/etsi_en/303700_303799/303798/02.01.01_60/en_303798v020101p.pdf
11. ETSI TS 103 175 V1.1.1 (2015–06): Intelligent Transport Systems (ITS); Cross Layer DCC Management Entity for operation in the ITS G5A and ITS G5B medium, available at https://www.etsi.org/deliver/etsi_ts/103100_103199/103175/01.01.01_60/ts_103175v010101p.pdf
12. Draft COMMISSION DELEGATED REGULATION (EU) …/... of XXX supplementing Directive 2010/40/EU of the European Parliament and of the Council with regard to the deployment and operational use of cooperative intelligent transport systems, available at https://www.europarl.europa.eu/cmsdata/161226/Delegated%20Regulation%20C-ITS.pdf
13. 5GAA welcomes Council objection against C-ITS Delegated Act, 5G Automotive Association, available at https://5gaa.org/5gaa-welcomes-council-objection-against-c-its-delegated-act/
14. FCC TO VOTE ON AUTO SAFETY SPECTRUM RULES, FCC Press Release, available at https://docs.fcc.gov/public/attachments/DOC-404027A1.pdf
15. C-V2X Standardisation in China, 5G Automotive Association, available at https://5gaa.org/content/uploads/2022/10/C-V2X-standardisation-in-China.pdf
16. Report from CEPT to the European Commission in response to the Mandate to study the extension of the Intelligent Transport Systems (ITS) safety-related band at 5.9 GHz, CEPT Report 71, March 2019, available at https://docdb.cept.org/download/126
17. Directive (EU) 2023/2661 of the European Parliament and of the Council of 22 November 2023 amending Directive 2010/40/EU on the framework for the deployment of Intelligent Transport Systems in the field of road transport and for interfaces with other modes of transport (Text with EEA relevance), available at https://eur-lex.europa.eu/eli/dir/2023/2661/oj
18. Hamida EB, Noura H, Znaidi W. Security of Cooperative Intelligent Transport Systems: Standards, Threats Analysis and Cryptographic Countermeasures. Electronics. 2015; 4(3):380–423. https://doi.org/10.3390/electronics4030380
19. M. Hasan, S. Mohan, T. Shimizu and H. Lu, "Securing Vehicle-to-Everything (V2X) Communication Platforms," in IEEE Transactions on Intelligent Vehicles, vol. 5, no. 4, pp. 693–713, Dec. 2020. https://doi.org/10.1109/TIV.2020.2987430
20. A. Czyżewski, A. Sroczyński, T. Śmiałkowski and P. Hoffmann, "Development of Intelligent Road Signs with V2X Interface for Adaptive Traffic Controlling," 2019 6th International Conference on Models and Technologies for Intelligent Transportation Systems (MT-ITS), Cracow, Poland, 2019, pp. 1–7, https://doi.org/10.1109/MTITS.2019.8883369
21. Liu Z, Liu Z, Meng Z, Yang X, Pu L, Zhang L. Implementation and performance measurement of a V2X communication system for vehicle and pedestrian safety. International Journal of Distributed Sensor Networks. 2016;12(9). https://doi.org/10.1177/1550147716671267

22. Chen, S., Hu, J., Shi, Y., Peng, Y., Fang, J., Zhao, R., & Zhao, L. (2017). Vehicle-to-everything (V2X) services supported by LTE-based systems and 5G. IEEE Communications Standards Magazine, 1(2), 70–76. https://doi.org/10.1109/MCOMSTD.2017.1700015
23. Ghosal, A., Conti, M. (2020). Security issues and challenges in V2X: A survey. Computer Networks, 169, 107093. https://doi.org/10.1016/j.comnet.2019.107093
24. Takahito Yoshizawa, Dave Singelée, Jan Tobias Muehlberg, Stephane Delbruel, Amir Taherkordi, Danny Hughes, and Bart Preneel. 2023. A Survey of Security and Privacy Issues in V2X Communication Systems. ACM Comput. Surv. 55, 9, Article 185 (September 2023), 36 pages. https://doi.org/10.1145/3558052
25. COMMISSION DECISION of 15 February 2011 concerning the adoption of the Working Programme on the implementation of Directive 2010/40/EU, February 2011, available at https://transport.ec.europa.eu/document/download/5f12b337-4abf-4028-852e-0cfc1f2791de_en?filename=c_2011_0289_en.pdf
26. COMMISSION DECISION of 11.12.2018 updating the Working Programme in relation to the actions under Article 6(3) of Directive 2010/40/EU, December 2018, available at https://transport.ec.europa.eu/document/download/5833f051-e82b-47fa-9c1b-961d2248b45d_en?filename=c20188264_en.pdf
27. COMMISSION DECISION of 13.12.2022 updating the working programme in relation to the actions under Article 6(3) of Directive 2010/40/EU for the period 2022–2027, December 2022, available at https://transport.ec.europa.eu/document/download/0ba5194e-429e-4065-9c7d-9faadbbb96e8_en?filename=C_2022_9140.pdf
28. Public workshop ITS Working Programme 2024–2028, European Commission, March 2024, available at https://transport.ec.europa.eu/document/download/434230ea-5bdc-4025-a5da-58c3d15f427d_en?filename=Programme_AW_2024.pdf&prefLang=ga
29. Andreas Festag, Cooperative intelligent transport systems standards in Europe, December 2014, IEEE Communications Magazine 52(12):166–172, 52(12):166–172, https://doi.org/10.1109/MCOM.2014.6979970
30. Autonomous Vehicles & the Role of C-V2X Cellular Technology, Hemant Rawat, August 2023, available at https://www.iotforall.com/autonomous-vehicles-and-the-role-of-5g-cellular-technology
31. ISO 26262:2018 – Road Vehicles Functional Safety Standards, available at https://webstore.ansi.org/standards/iso/iso26262roadvehiclesfunctional?source=blog&_gl=1*qkvmc*_gcl_au*NTQwMjczODc4LjE3MjUyMDk2MjQ.
32. CV 265: Introduction to IEEE 1609 Family of Standards for Wireless Access in a Vehicular Environments (WAVE), available at https://www.its.dot.gov/StandardsTraining/mod63/sup/m63sup.pdf
33. Connected Vehicle Deployment, Technical Assistance, Security Credential Management System (SCMS) Technical Primer, Report – November 2019, FHWA-JPO-19–775, available at file:///C:/Users/mumarkus/Downloads/dot_43635_DS1.pdf
34. ETSI TS 102 941 V2.2.1 (2022–11) Intelligent Transport Systems (ITS); Security; Trust and Privacy Management; Release 2, available at https://www.etsi.org/deliver/etsi_ts/102900_102999/102941/02.02.01_60/ts_102941v020201p.pdf
35. Shanzhi Chen, Jinling Hu, Li Zhao, Rui Zhao, Jiayi Fang, Yan Shi, Hui Xu, Cellular Vehicle-toEverything (C-V2X), Posts & Telecom Press and Springer, available at https://link.springer.com/book/10.1007/978-981-19-5130-5
36. 5.9 GHz band configuration for road-ITS deployment in Europe, 5G Automotive Association, March 2024, available at https://5gaa.org/content/uploads/2024/03/5gaa-position-paper-european-road-its-deployment.v3.pdf
37. ECC Recommendation (08)01: Use of the band 5855–5875 MHz for Intelligent Transport Systems (ITS), Approved 21 February 2008, Amended 3 July 2015, CEPT ECC, available at https://docdb.cept.org/download/3039
38. ECC Decision (08)01: The harmonised use of Safety-Related Intelligent Transport Systems (ITS) in the 5875–5935 MHz frequency band, approved 14 March 2008, latest updated 18 November 2022, CEPT ECC, available at https://docdb.cept.org/download/4232

39. ETSI EN 300 674–2–1 V3.1.1 (2022–03), Transport and Traffic Telematics (TTT); Dedicated Short Range Communication (DSRC) transmission equipment (500 kbit/s / 250 kbit/s) operating in the 5 795 MHz to 5 815 MHz frequency band; Part 2: Harmonised Standard for access to radio spectrum; Sub-part 1: Road Side Units (RSU)
40. ETSI EN 300 674–2–2 V2.2.1 (2019–06), Transport and Traffic Telematics (TTT); Dedicated Short Range Communication (DSRC) transmission equipment (500 kbit/s / 250 kbit/s) operating in the 5 795 MHz to 5 815 MHz frequency band; Part 2: Harmonised Standard for access to radio spectrum; Sub-part 2: On-Board Units (OBU)
41. ECC News March 2024, CEPT ECC, available at https://cept.org/ecc/news/ecc-news-march-2024
42. IEEE 1609 - Family of Standards for Wireless Access in Vehicular Environments (WAVE), available at https://www.standards.its.dot.gov/Factsheets/Factsheet/80
43. ETSI EN 302 663 V1.3.1 (2020–01) Intelligent Transport Systems (ITS); ITS-G5 Access layer specification for Intelligent Transport Systems operating in the 5 GHz frequency band, available at https://www.etsi.org/deliver/etsi_en/302600_302699/302663/01.03.01_60/en_302663v010301p.pdf
44. 3GPP TS 33.185, Security aspect for LTE support of Vehicle-to-Everything (V2X) services, available at https://portal.3gpp.org/desktopmodules/Specifications/SpecificationDetails.aspx?specificationId=3141
45. 3GPP TS 33.356, Security aspects of 3GPP support for advanced Vehicle-to-Everything (V2X) services, available at https://portal.3gpp.org/desktopmodules/Specifications/SpecificationDetails.aspx?specificationId=3724
46. 3GPP TS 33.401, 3GPP System Architecture Evolution (SAE); Security architecture, available at https://portal.3gpp.org/desktopmodules/Specifications/SpecificationDetails.aspx?specificationId=2296
47. Commission Delegated Regulation (EU) No 305/2013 of 26 November 2012 supplementing Directive 2010/40/EU of the European Parliament and of the Council with regard to the harmonised provision for an interoperable EU-wide eCall Text with EEA relevance, available at https://eur-lex.europa.eu/legal-content/EN/TXT/?uri=CELEX%3A32013R0305
48. Commission Delegated Regulation (EU) 2024/1084 of 6 February 2024 amending Delegated Regulation (EU) No 305/2013 supplementing Directive 2010/40/EU of the European Parliament and of the Council with regard to the harmonised provision for an interoperable EU-wide eCall, available at https://eur-lex.europa.eu/legal-content/EN/TXT/?uri=OJ%3AL_202401084
49. Commission Delegated Regulation (EU) No 886/2013 of 15 May 2013 supplementing Directive 2010/40/EU of the European Parliament and of the Council with regard to data and procedures for the provision, where possible, of road safety-related minimum universal traffic information free of charge to users Text with EEA relevance, available at https://eur-lex.europa.eu/legal-content/EN/TXT/?uri=CELEX%3A32013R0886
50. Commission Delegated Regulation (EU) No 885/2013 of 15 May 2013 supplementing ITS Directive 2010/40/EU of the European Parliament and of the Council with regard to the provision of information services for safe and secure parking places for trucks and commercial vehicles Text with EEA relevance, available at https://eur-lex.europa.eu/legal-content/EN/TXT/?uri=CELEX%3A32013R0885
51. Commission Delegated Regulation (EU) 2015/962 of 18 December 2014 supplementing Directive 2010/40/EU of the European Parliament and of the Council with regard to the provision of EU-wide real-time traffic information services (Text with EEA relevance), available at https://eur-lex.europa.eu/legal-content/EN/TXT/?uri=CELEX%3A32015R0962
52. Commission Delegated Regulation (EU) 2017/1926 of 31 May 2017 supplementing Directive 2010/40/EU of the European Parliament and of the Council with regard to the provision of EU-wide multimodal travel information services (Text with EEA relevance), available at https://eur-lex.europa.eu/legal-content/EN/TXT/?uri=CELEX%3A32017R1926
53. Commission Delegated Regulation (EU) 2024/490 of 29 November 2023 amending Delegated Regulation (EU) 2017/1926 supplementing Directive 2010/40/EU of the European Parliament and of the Council with regard to the provision of EU-wide multimodal travel information

services, available at https://eur-lex.europa.eu/legal-content/EN/TXT/?uri=CELEX%3A32024R0490

54. Commission Delegated Regulation (EU) 2022/670 of 2 February 2022 supplementing Directive 2010/40/EU of the European Parliament and of the Council with regard to the provision of EU-wide real-time traffic information services (Text with EEA relevance), available https://eur-lex.europa.eu/legal-content/EN/TXT/?uri=CELEX%3A32022R0670

Chapter 7
Introduction to the Networks and Information Systems 2 (NIS2) Directive

Christophe Gaie and Markus Mueck

Abstract The Networks and Information Systems 2 (NIS2) Directive has been published in 2022. It is the successor of its initial version (NIS Directive) of 2016. Its objective is to achieve a high common level of cybersecurity across the European Union, with a view to improving the functioning of the internal market. The previously existing legal framework was indeed revised to better address the increased digitization and an evolving cybersecurity threat landscape. New sectors and entities have been added to the scope of the regulation to improves the resilience and incident response capacities of public and private entities, competent authorities and the European Union as a whole. For this purpose, the NIS2 Directive introduces a list of "Sectors of high criticality" including Energy, Transport, Banking, Financial Market Infrastructures, Health, Drinking Water, Waste Water, Digital Infrastructure, ICT Service Management (Business-to-Business), Public Administration and Space. A set of measures is introduced for those sectors which will be further summarized in the present chapter. Additionally, with requirements of a lesser extent, "other critical Sectors" are defined including Postal and Courier Services, Waste Management, Manufacture, production and distribution of chemicals, Production, processing and distribution of food, Manufacturing, Digital Providers and Research. The application of the NIS2 Directive is limited to public or private entities being part of the list of "Sectors of high criticality" or the list of "other critical Sectors". The requirements of the NIS2 Directive remain on a rather high level with the obligation for EU Member States to detail the implementation on a national level. According to NIS2 Article 41, by 17 October 2024, Member States shall adopt and publish the measures necessary to comply with the NIS2 Directive.

Keywords Regulation · Digital · Cyber security · Market access · European Union Single Market

---

C. Gaie (✉)
Head of the Engineering and Digital Innovation Division, Prime Minister Services, Paris, France
e-mail: Christophe.Gaie@gmail.com

M. Mueck
Intel Deutschland GmbH, Neubiberg, Germany
e-mail: Markus.Dominik.Mueck@intel.com

M. Mueck and C. Gaie (eds.), *European Digital Regulations*, Intelligent Systems Reference Library 265, https://doi.org/10.1007/978-3-031-80809-8_7

7.1 Introduction

7.1.1 A Brief History of Cybersecurity

Cybersecurity emerged in the 1970s with the advent of ARPANET. The rapid expansion of computer networks since then explains the growth in cyber threats and vulnerabilities. To ensure the protection of machines, people and organization, many security solutions were imagined and commercialized by the end of the twentieth century [1].

The 1980s saw the emergence of the first antivirus solutions to combat the growing threat of computer viruses [2]. Early antivirus programs, often developed by individuals or small groups, focused on detecting and removing simple virus signatures. As computer viruses became more sophisticated, antivirus software evolved to include features such as heuristic analysis and real-time protection.

The widespread adoption of the internet in the 1990s led to a significant increase in cyber threats. New types of malware, such as worms, trojans, and spyware, emerged, exploiting vulnerabilities in internet protocols and applications [3]. Additionally, the rise of online commerce and the exchange of sensitive data created new opportunities for cybercriminals to steal personal information and financial data.

The twenty-first century brought sophisticated cyberattacks, fueled by organized crime and state-sponsored actors. The cybersecurity industry has responded with rapid growth, but the evolving threat landscape, exemplified by the rise of ransomware, demands continuous innovation and adaptation. The heterogeneity of responses does not guarantee effective and optimal protection, which helps to justify the need to implement regulation at the EU level [4]. This is illustrated in Fig. 7.1.

Cybersecurity is defined as the domain that ensures the protection of digital systems, networks, and data from unauthorized access, use, disclosure, disruption, modification, or destruction. It encompasses protecting the confidentiality, integrity, and availability of information. As our reliance on technology grows, so does the complexity and sophistication of cyber threats, making cybersecurity a critical concern for individuals, businesses, and governments alike [5].

7.1.2 Notable Cyberattacks and Their Implications

Understanding the efforts towards regulating and protecting strategic IT requires to have a certain knowledge of the risks encountered and existing threats. Thus, the authors propose a brief overview of some significant cybersecurity incidents to highlight the escalating nature of these threats and their far-reaching consequences. The timeline proposed in Fig. 7.2 illustrates the different steps of evolution of cyberattacks [6].

One of the earliest and most notable computer worms, the Morris Worm (1988), demonstrated the potential for self-replicating malware to spread rapidly across

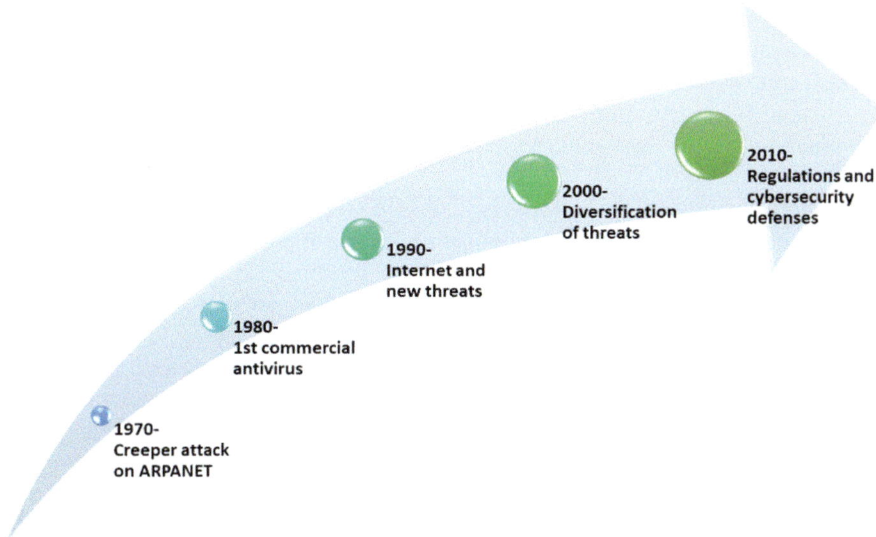

Fig. 7.1 A brief history of cybersecurity

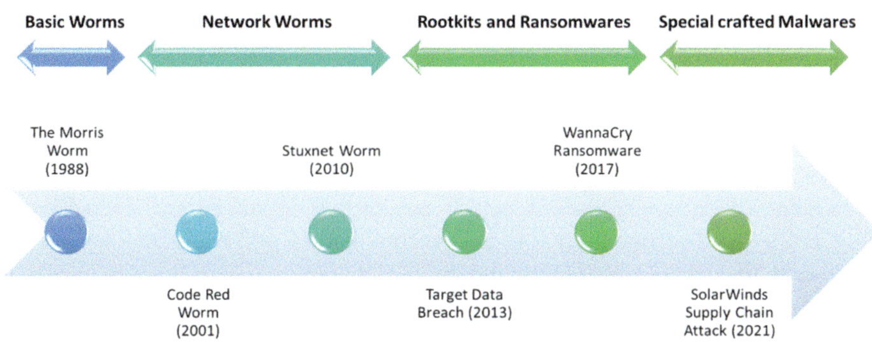

Fig. 7.2 Different steps of cyberattacks

interconnected networks [7]. This incident underscored the vulnerability of nascent computer systems and the need for robust security measures.

Then, Code Red worm infected hundreds of thousands of web servers in 2001, causing devastating widespread worm infestations. This attack highlighted the importance of patching vulnerabilities promptly and implementing effective network security measures to prevent such large-scale outbreaks.

The Stuxnet worm, believed to be a joint US-Israeli operation, targeted Iran's nuclear program in 2010. This sophisticated attack demonstrated the potential for nation-states to use cyber weapons to achieve geopolitical objectives. It also raised concerns about the growing sophistication of cyber threats and the need for robust defenses against state-sponsored attacks.

The Target data breach, one of the largest retail data breaches in history, exposed the personal information of millions of customers in 2013. This incident exposed the vulnerabilities of large-scale retail systems and the severe consequences of data breaches, including financial losses, reputational damage, and regulatory fines.

WannaCry ransomware attack, which affected hundreds of thousands of computers worldwide in 2017, highlighted the global impact of ransomware attacks. By encrypting files and demanding a ransom for their decryption, ransomware attacks can disrupt critical services, cause financial losses, and damage organizational reputation.

The SolarWinds supply chain attack, which compromised numerous government and private sector organizations in 2021, exposed the vulnerabilities of software supply chains. By infiltrating a trusted software vendor, attackers were able to gain access to a wide range of targets, demonstrating the potential for sophisticated supply chain attacks to cause significant damage.

The authors conclude that the increasing sophistication of cyber threats and the growing reliance on digital technologies make it imperative for individuals, businesses, and governments to take proactive measures to protect themselves from these risks. This includes implementing strong cybersecurity measures, educating users about safe online practices, and fostering international cooperation to address global cyber threats. However, voluntary measures alone may not be sufficient to ensure adequate cybersecurity protections, especially in critical sectors like healthcare, finance, and energy.

Thus, by establishing clear cybersecurity regulations, governments can create a more secure and resilient digital environment for individuals, businesses, and society as a whole. These regulations can help to ensure that organizations have adequate cybersecurity measures in place, promote accountability, and facilitate international cooperation to address global cyber threats.

7.1.3 A Brief History of Cybersecurity

This section proposes an overview that highlights the evolution of cybersecurity to provide a better protection of digital assets to the benefits of companies, citizens and government services. As the threat landscape evolves and becomes more sophisticated, robust cybersecurity measures have been optimized to mitigate risks and safeguard sensitive information [8]. Figure 7.3 and the paragraphs below illustrate this evolution (awareness, understanding, implementation, expansion, expertise).

One of the first publication concerning information security and privacy was achieved by Willis Ware at the spring joint computer conference [9]. It outlines the necessity to define measures to avoid the leakage of information. During this decade, the first computer security models were also developed. These models, such as the Bell-LaPadula model, provided a theoretical framework for implementing security

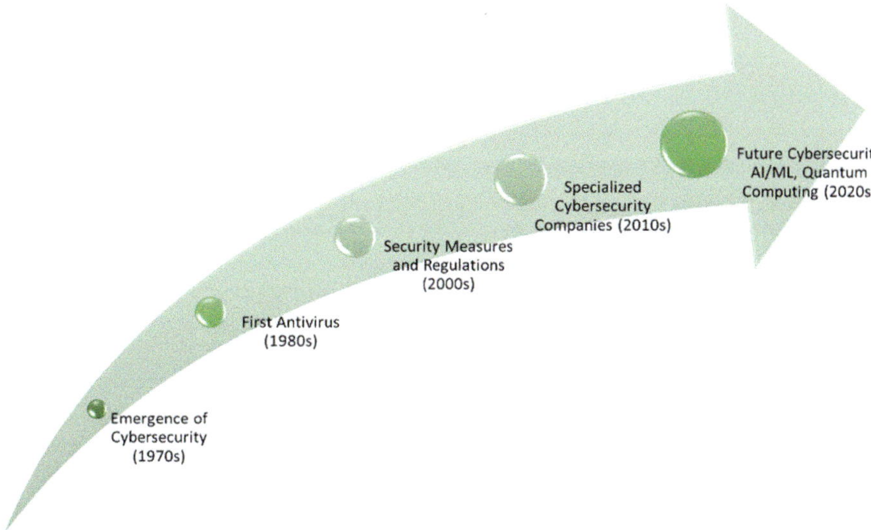

Fig. 7.3 Evolution and progress of cybersecurity

controls in computer systems [10]. These models are still used today to develop cybersecurity policies and procedures.

The 1970s also saw the development of the first computer security standards. The United States Department of Defense published the "Orange Book" in 1978, which provided guidelines for computer security [11]. This book was part of a series known as the Rainbow Series, which covered various aspects of computer security.

The rise of internet connectivity led to a significant increase in personal computers in the following decade. This growth brought new challenges, including the proliferation of computer viruses and malware. The development of antivirus software, such as Ultra Virus Killer or McAfee, offered early protection against these threats. However, the increasing sophistication of cyberattacks, including the Morris Worm, emphasized the need for more comprehensive security measures.

Then, major data breaches and malwares were revealed during the 2000s. As a consequence, governments and organizations began to recognize the potential for cyberattacks to disrupt critical infrastructure and steal sensitive information. The Stuxnet worm, designed to target Iranian nuclear facilities, showcased the destructive capabilities of state-sponsored cyberattacks. Meanwhile, high-profile data breaches exposed millions of personal records, highlighting the vulnerability of digital systems.

Since the 2010s, many cybersecurity companies were created and offer a wide range of services to protect organizations from cyber threats. The growing complexity of attacks and the increasing reliance on digital infrastructure fueled the demand for specialized cybersecurity expertise. Advanced technologies, such as artificial

Table 7.1 Recent cybersecurity solutions and their usage

Solution	Usage
Endpoint Detection and Response (EDR)	For protection and response capabilities
Extended Detection and Response (XDR)	For consolidation responses and information
Security Information and Event Management (SIEM)	For centralization of information, reporting and multi-IT systems
Managed Detection and Response (MDR)	For proactive detection and incident response
Security Orchestration, Automation, and Response (SOAR)	For incident response automation and better information sharing between IT teams

intelligence (AI) and machine learning (ML), emerged as powerful tools for threat detection and prevention [12, 13].

The recent research highlighted the urgent need for organizations to adopt a proactive and comprehensive approach to cybersecurity. Another key dimension is to articulate the different technologies and understand when employing them as described in Table 7.1.

7.1.4 The Network and Information Security (NIS) Directive

As indicated previously, the decentralized nature of cyberspace made it challenging to establish a unified approach to cybersecurity. The successive phases below illustrate the progressive evolution towards the establishment of the NIS Directive.

In 2013, the EU adopted its first cybersecurity strategy, [14] aiming for an "Open, Safe and Secure Cyberspace" as indicated in [15]. This strategy outlined five key objectives (Fig. 7.4).

To address cybercrime, the strategy promoted cooperation between the newly formed European Cybercrime Centre (EC3) and fostered collaboration between public and private sectors to strengthen defense capabilities.

Then, the EU Cyber Defense Policy Framework was adopted in 2014 and aims to define the cyber defense aspects of the EU Cybersecurity Strategy [16]. It sets up the following cyber defense priorities (Fig. 7.5).

Furthermore, it also clarifies the roles of the different European actors to guarantee the cybersecurity within the European Union.

Finally, in 2015, the European Commission adopted the Digital Single Market (DSM) strategy, a multifaceted package of initiatives that promotes multiple initiatives towards data security and cyber protection [17]. The emphasis on personal data appears in initiative 12, that aims to build the EU's data protection rules, eventually resulting in the General Data Protection Regulation (GDPR). Moreover, initiative 13 outlines the importance of cybersecurity and promotes a partnership between the Commission and industry in the domain of cybersecurity (Fig. 7.6).

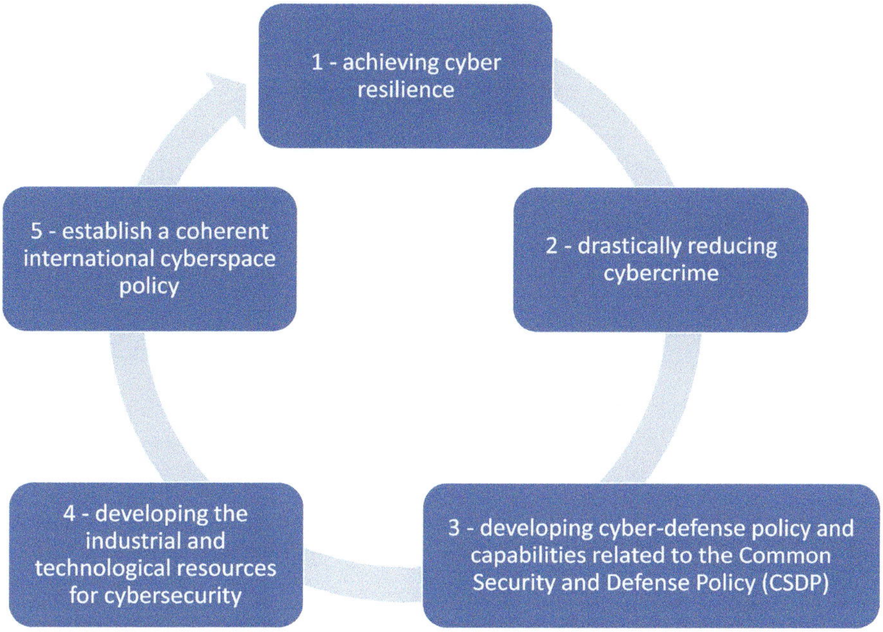

Fig. 7.4 Objectives of the first EU cybersecurity strategy of 2013

The Network and Information Security (NIS) Directive marked a significant step, linking cybersecurity to critical infrastructure protection [18]. This Directive introduced a multi-layered approach, imposing obligations on various levels.

The EU was tasked with establishing a cooperation group to facilitate information sharing among member states and creating a network of computer security incident response teams (CSIRTs). Member states were required to develop national cybersecurity strategies, designate competent authorities, and establish at least one CSIRT. Operators of essential services (OESs) and digital service providers (DSPs) were subject to specific security requirements.

What is a CSIRT?

A Computer Security Incident Response Team (CSIRT) is the entity in charge of tracking and addressing reported cybersecurity incidents. In addition to responding to incidents, the CSIRT issues warnings, alerts, and announcements about potential threats. Relevant authorities are obligated to promptly inform the CSIRT of any cybersecurity incidents they encounter

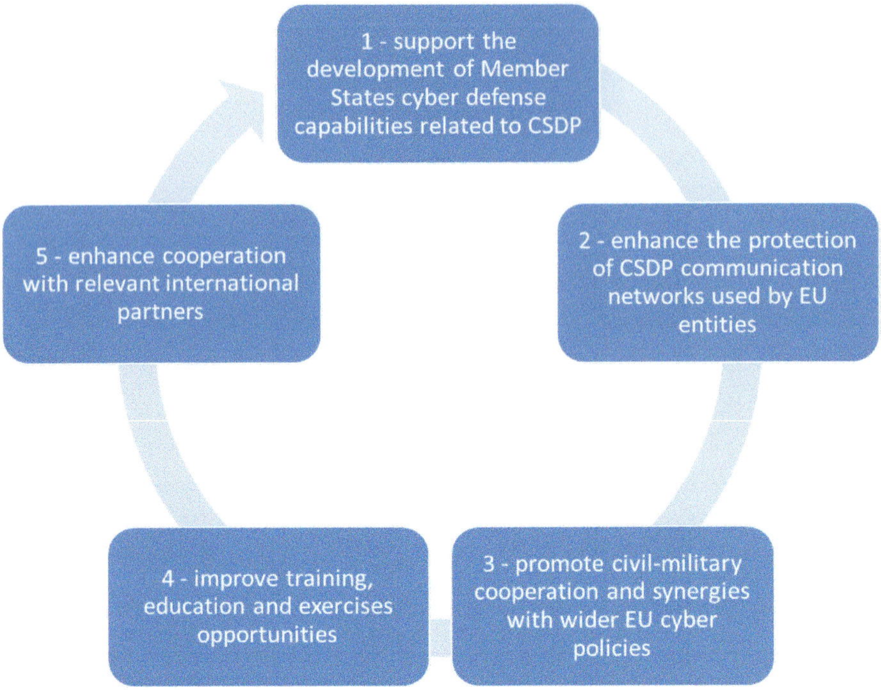

Fig. 7.5 Cyber defense priorities of the EU Cyber Defense Policy Framework [16]

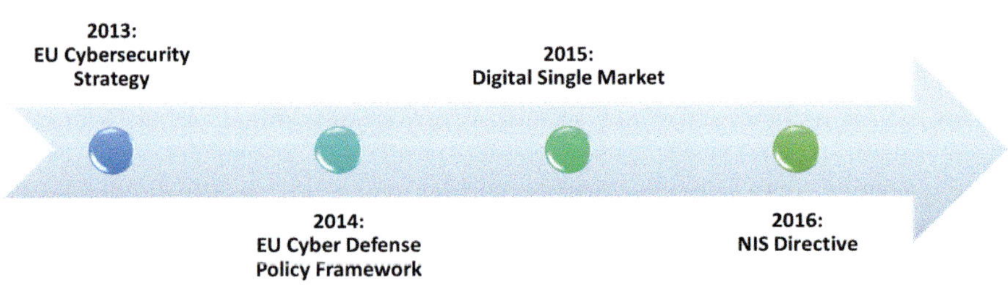

Fig. 7.6 Timeline towards the NIS Directive

7.2 General Overview of NIS2

7.2.1 Why NIS2 Was Required?

NIS2 (Network and Information Systems Directive 2) was introduced due to the increasing complexity and severity of cyber threats [19]. The original NIS Directive, while a step in the right direction, was deemed insufficient to address the evolving landscape. The following motivations explain why it is crucial to implement NIS2:

- **Rising Cyber Threats:** The frequency and impact of cyberattacks have escalated dramatically. Critical infrastructure and essential services are increasingly targeted, making robust cybersecurity measures imperative.
- **Economic Impact:** Cyberattacks can cause significant financial losses, disrupt supply chains, and damage consumer confidence. NIS2 aims to protect the EU's economy by enhancing cybersecurity resilience.
- **Harmonization:** The previous NIS Directive lacked uniformity across EU member states. NIS2 seeks to establish a common level of cybersecurity protection throughout the EU, creating a more secure digital environment.
- **Expanded Scope:** NIS2 covers a wider range of organizations, including digital service providers, to ensure a more comprehensive approach to cybersecurity. The Directive explicitly extends the scope to essential and important entities in its Article 3.
- **Strengthened Enforcement:** With stricter penalties for non-compliance, NIS2 encourages organizations to prioritize cybersecurity. Member states should now verify that administrative fines imposed in case of infringements of this Directive are effective, proportionate and dissuasive. This topic is detailed in Article 34 of the NIS2 Directive.
- **Incident Reporting:** Improved incident reporting mechanisms allow for faster response and better understanding of emerging threats. Each member state must set up a CSIRT (as detailed in Article 12) that interacts with the CSIRT network (as explained in Article 15).
- **Supply Chain Security:** Recognizing the vulnerability of supply chains, NIS2 addresses the need to protect against attacks originating from third-party suppliers. The importance of managing supply chains is details in Article 21 and Article 22 of the Directive. It is also important to underline the creation of the European cyber crisis liaison organisation network (EU-CyCLONe) to support the coordinated management of large-scale cybersecurity incidents and crises at operational level (as described in Article 16) (Fig. 7.7).

7.2.2 What Was the Adoption Process?

The initiation of the NIS2 legislative process begins by the recognition of shortcomings in NIS1 as detailed in the previous sections. The limits were observed during the progressive application of NIS1 between the adoption of the Directive (July 6, 2016) and its transposition deadline (May 9, 2018).

Then, the legislative process was applied to define the contours of the new Directive. First, the commission proposed a new Directive on 16 December 2020. After 18 months of work and evolutions of the project, it was validated by the Industry, Research and Energy (ITRE) committee on 13 July 2022 and by Parliament on 10 November 2022. Finally, the NIS2 Directive was definitely adopted on 14 December 2022. Member States have 21 months after the entry into force of the Directive to transpose it into national law [20].

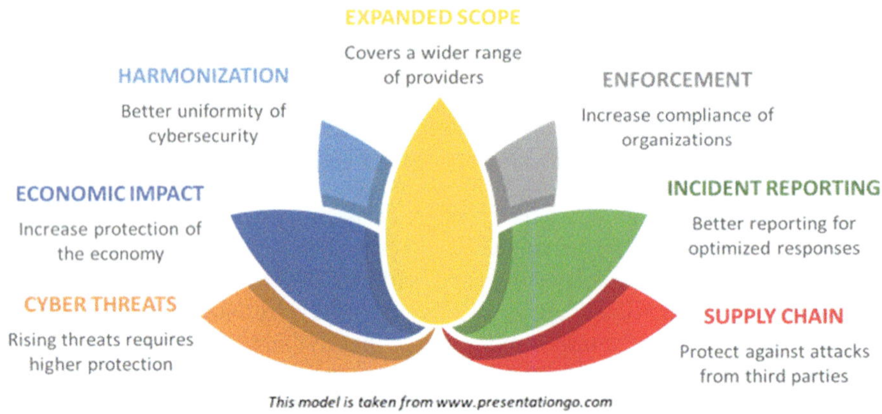

Fig. 7.7 Motivations to establish the NIS2 Directive

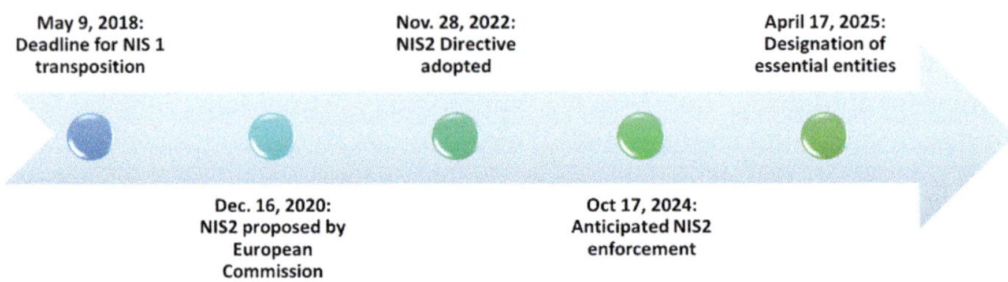

Fig. 7.8 Timeline for the enforcement of the NIS2 Directive

According to NIS2 Article 41, European Union Member States are required to adopt and publish appropriate measures to comply with NIS2 by October 17th, 2024. On the following day, October 18th, 2024, the measure shall apply.

At the time when the present chapter was written, Germany had published a draft legislation laying out the transposition of the NIS2 Directive for Germany [21].

The following figure illustrates the timeline of the adoption process (Fig. 7.8).

7.3 Impact of NIS2 on Businesses

7.3.1 Overview

The Networks and Information Systems 2 (NIS2) Directive is designed to provide a high-level framework and related objectives, together with a set of minimum requirements, that are handed over to European Union Member States, including a request

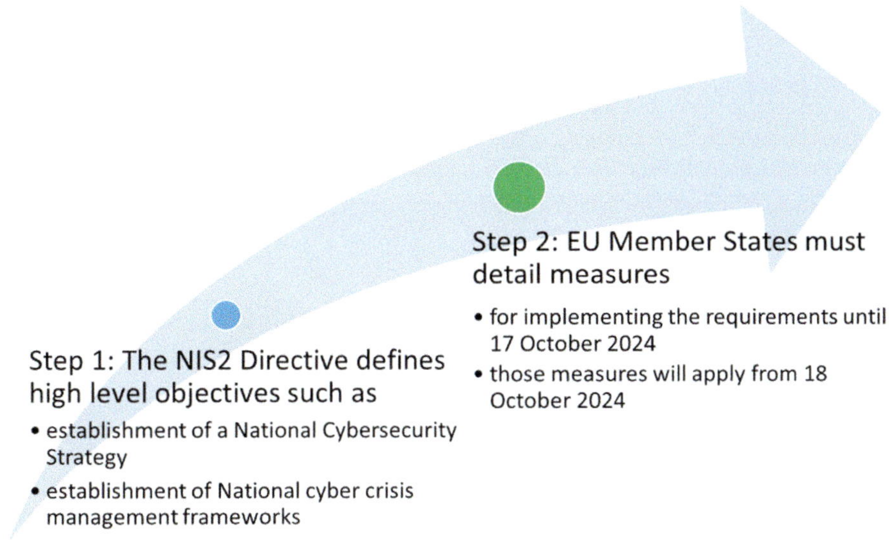

Fig. 7.9 High level approach of the NIS2 Directive

for setting up specific activities and institutions which will be discussed later in the present section. The overall approach is illustrated below (Fig. 7.9).

As defined in Article 2 of the NIS2 Directive, the requirements of the NIS2 Directive apply to public or private entities (which qualify as medium-sized enterprises as defined by [22] or exceed the ceilings for medium-sized enterprises and which provide their services or carry out their activities within the Union) being part of the list of "Sectors of high criticality" or the list of "other critical Sectors" which are defined in Annex I and Annex II of NIS2 respectively as follows (Fig. 7.10).

Moreover, whatever their size, the Directive also applies to stakeholders of specific importance as further detailed in Article 2(2), Article 2(3) and Article 2(4) of NIS2; those include providers of public electronic communications networks (for example mobile and fixed network providers), trust service providers (for example related to digital signatures), top-level domain name registries (internet top-level domains include.com,.org, etc.) and domain name system service providers (as described in Article 2). It also concerns any essential provider of a Member State that ensures the maintenance of critical societal or economic activities (such as electricity production and energy storage, management and maintenance of airport or railways infrastructure, etc.), entities whose disruption could have a significant impact on public safety, public security or public health as well as a systemic risk, entities identified as critical entities as defined in [23] and entities providing domain name registration services. Finally, public administration entities of national level have to apply the Directive as well as regional ones on a risk-based assessment for the latter. Further details on the applicability of NIS2 are defined in Article 2(6) to Article 2(14).

Fig. 7.10 "Sectors of high criticality" and "other critical Sectors" as defined by the NIS2 Directive in Annex I and Annex II respectively

Within the framework outlined above, NIS2 lays down the following requirements (as defined by its Article 1(2)) (Fig. 7.11).

The Directive also categorizes organizations into two main groups:

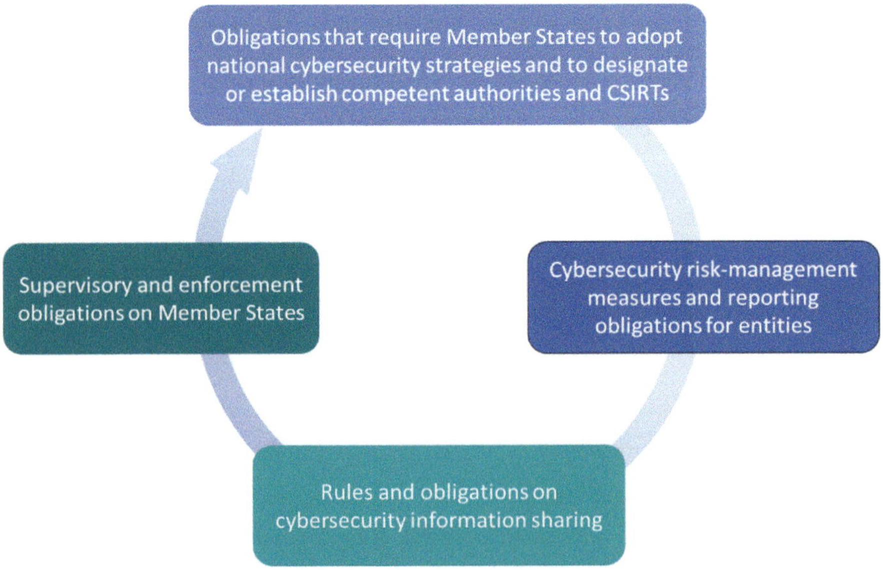

Fig. 7.11 Overview on requirements by the NIS2 Directive

- **Essential Entities (EEs)**: These are organizations critical to the functioning of society and the economy. They are subject to more stringent requirements. Examples include energy, transport, banking, health, digital infrastructure, and water supply sectors.
- **Important Entities (IEs)**: These entities have a significant impact on the provision of essential services. They are subject to less stringent requirements compared to EEs. Examples include food, chemicals, manufacturing, and digital service providers meeting certain size criteria.

It's crucial to note that the specific definition of EEs and IEs can vary between member states.

To enforce the cybersecurity objectives, NIS2 introduces a number of requirements on Member States as follows (Fig. 7.12).

A crucial aspect of the NIS2 Directive is the requirement for each Member State to establish a "National Cybersecurity Strategy" (Article 7). This strategy must outline strategic objectives, necessary resources, and appropriate policy and regulatory measures to attain and sustain a high level of cybersecurity.

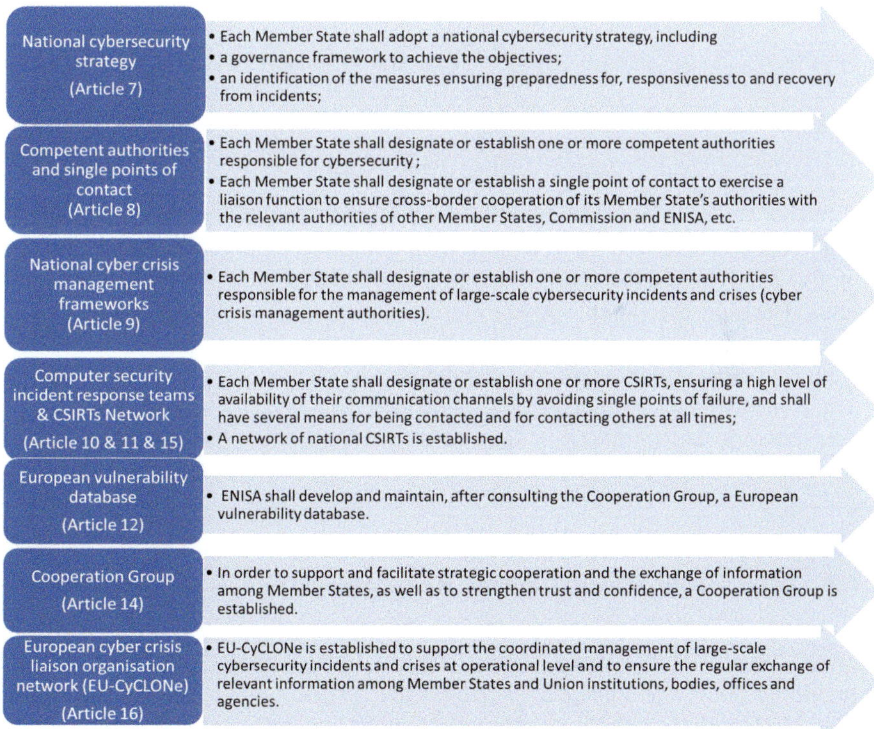

Fig. 7.12 Summary of requirements by the NIS2 Directive

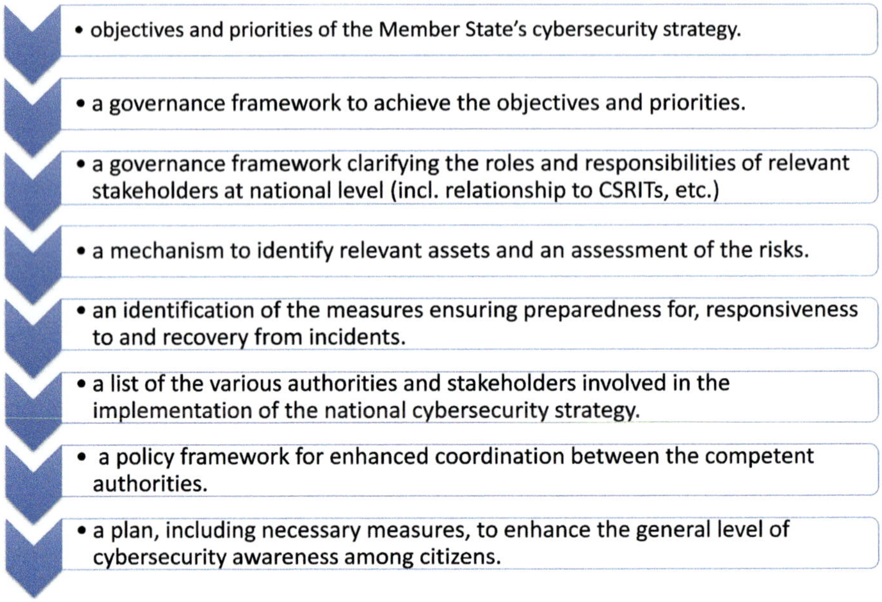

Fig. 7.13 Required elements of a National Cybersecurity Strategy

Selected key elements of a National Cybersecurity Strategy are summarized below in a simplified way (Fig. 7.13).

A further key component of the NIS2 Directive is Article 21: Cybersecurity risk-management measures where … technical, operational and organisational measures are defined to manage the risks posed to the security of network and information systems. The objective is to prevent or minimise the impact of incidents on recipients of their services. Those measures should … build on the state-of-the-art and … relevant European and international standards.

The measures outlined above need to be based on an … approach that aims to protect network and information systems and the physical environment …. At a minimum, those measures must include the following (Fig. 7.14).

7.3.2 *Interrelations with Other Regulations*

The NIS2 Directive [19] references the European Cybersecurity Act [1] in its Article 24 regarding the use of European cybersecurity certification schemes. Under NIS2 Article 21, Member States may require essential and important entities to utilize specific ICT products, services, or processes, either developed internally or procured from third parties, that are certified under European cybersecurity certification

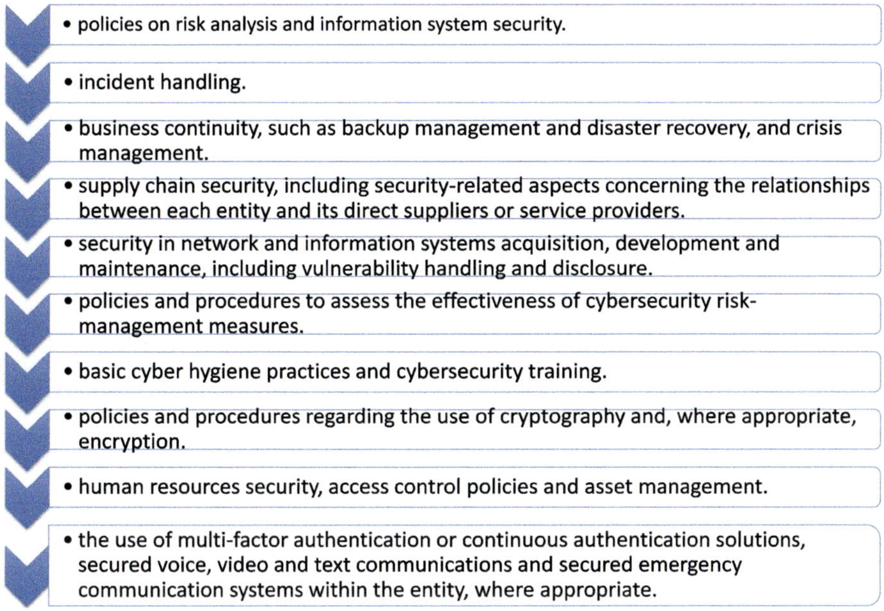

Fig. 7.14 Minimum measures for compliance with Article 21: Cybersecurity risk-management measures of the NIS2 Directive

schemes established by the European Cybersecurity Act [1]. Moreover, if inadequate cybersecurity levels are detected, the European Commission has the authority to mandate the use of certain certified ICT products, services, or processes or require entities to obtain a certificate under a European cybersecurity certification scheme through Delegated Acts.

7.3.3 Obligations to Fulfill and Impact of Non-compliance

There are multiple requirements to comply with the NIS2 Directive. Thus, Essential and Important Entities should take several measures to comply with the NIS2 Directive (as detailed in Chapter IV of the Directive): conducting regular risk assessments to identify vulnerabilities and implement mitigating measures, establishing incident response plans and promptly reporting cybersecurity incidents to competent authorities, assessing and managing cybersecurity risks within the supply chain, developing plans to ensure the continuity of critical services in case of a cyberattack, implementing training programs to raise cybersecurity awareness among employees, and cooperating with competent authorities and sharing information about cyber threats.

Non-compliance with NIS2 can result in significant penalties. The Directive sets minimum penalties, but member states can impose stricter sanctions (as detailed in Article 34). Essential Entities are subject to maximum fines of €10 million or 2% of global annual turnover, while Important Entities may face maximum fines of €7 million or 1.4% of global annual turnover. Additionally, national authorities can impose non-monetary penalties such as compliance orders or public reprimands.

7.4 Benefits of NIS2 Compliance

NIS2 compliance offers a comprehensive approach to enhancing an organization's cybersecurity posture. By mandating regular risk assessments, organizations gain the ability to proactively identify and address potential vulnerabilities before they can be exploited by malicious actors. This proactive stance is crucial in preventing costly data breaches and system failures, thereby safeguarding sensitive information and minimizing financial losses.

Effective incident management is another cornerstone of NIS2 compliance. Having well-defined incident response plans in place allows organizations to react swiftly and decisively when faced with cyberattacks. This rapid response capability helps limit the impact of incidents, preserving critical systems and data. Furthermore, demonstrating a proactive approach to incident management can significantly enhance an organization's reputation, fostering trust among customers, partners, and stakeholders.

Business continuity is a vital component of NIS2 compliance. By developing robust business continuity plans, organizations can ensure the resilience of their operations in the face of cyberattacks. These plans encompass various measures, including backup systems, disaster recovery strategies, and the maintenance of critical services. The ultimate goal is to minimize operational downtime and maintain customer satisfaction, even during challenging circumstances.

Supply chain security is another critical aspect addressed by NIS2. By assessing and managing cybersecurity risks within the supply chain, organizations can protect themselves from vulnerabilities that may arise from third-party systems. This proactive approach strengthens the overall security posture and reduces the risk of disruptions caused by cyberattacks on suppliers. A secure supply chain also contributes to an enhanced reputation, as it demonstrates a strong commitment to the protection of sensitive information.

NIS2 compliance can drive improved efficiency and productivity. Implementing robust cybersecurity measures can streamline operations, reduce operational costs, and empower employees to play an active role in identifying and reporting potential threats. Moreover, demonstrating a strong commitment to cybersecurity can position an organization as a trusted and reliable partner, providing a significant competitive advantage in today's digital landscape.

These paragraphs illustrate that NIS2 compliance offers a multitude of benefits that extend beyond mere regulatory compliance. By adopting a proactive approach

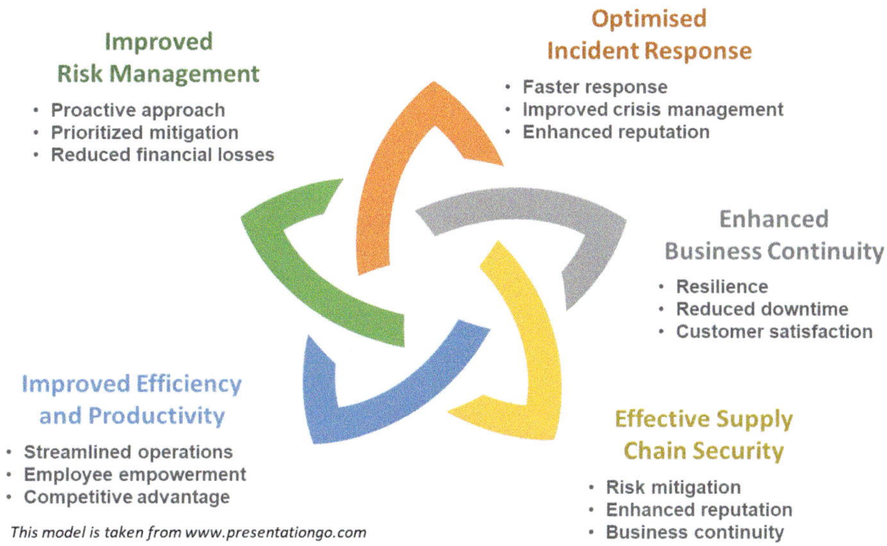

Fig. 7.15 Benefits of complying to the NIS2 Directive

to risk management, incident response, business continuity, supply chain security, and overall cybersecurity posture, organizations can significantly enhance their resilience, protect their reputation, and drive operational efficiency (Fig. 7.15).

7.5 Challenge of Compliance to NIS2

Achieving compliance with the NIS2 Directive presents significant challenges for organizations. The Directive covers a wide range of sectors and entities, making it difficult to develop a standardized approach. Additionally, the complexity of NIS2 requirements, including risk assessments, incident response plans, and supply chain security, can be overwhelming. The ever-evolving threat landscape further complicates matters, necessitating continuous adaptation of security measures. Lastly, implementing robust cybersecurity measures often requires substantial investments in technology, personnel, and processes, creating resource constraints for many organizations.

ISO 27001, a globally recognized standard for information security management systems, offers a valuable framework for achieving NIS2 compliance. By implementing ISO 27001, organizations can establish a structured approach to identifying, assessing, and managing information security risks. This aligns closely with the risk assessment requirements of NIS2. Moreover, ISO 27001 helps demonstrate a strong commitment to information security, building trust with customers, partners, and other stakeholders. Beyond NIS2, ISO 27001 also provides a foundation for complying with other cybersecurity regulations, such as the GDPR. However, it is important to underline that "compliance with ISO/IEC 27001 or NIS2 does not automatically mean an organization is compliant with the other" as detailed by Viheriäranta in [24].

Proactive preparation is essential for successful NIS2 compliance. Organizations should begin by thoroughly understanding the Directive's requirements and conducting a gap assessment to identify existing security controls and areas for improvement. Developing a phased implementation plan, involving key stakeholders, and staying informed about regulatory changes are crucial steps in the compliance journey. By investing in NIS2 compliance, organizations not only mitigate risks but also enhance their overall security posture, demonstrating a strong commitment to protecting sensitive information and building trust with stakeholders.

7.6 Conclusion

The Networks and Information Systems 2 (NIS2) will be key to achieve a high common level of cybersecurity across the European Union, with a view to improving the functioning of the internal market. The transposition deadline for EU Member States is set for 17 October 2024. As discussed above, Germany has made a draft legislative proposal [21] available when the current chapter was written and other Member States are expected to follow. The transposition of the directive is not in the same maturity level all around Europe [25] and some countries have already transposed the directive such as Belgium [26] and Hungary [27].

In the view of the authors of the present chapter, it will be important to ensure consistency of implementation over all EU member states. In case of divergence, there is the risk of excessive cost and effort required by industry to understand the legislative requirements across the entire European Union and to implement solutions meeting all of them. Still, there is no way around a horizontal approach to cybersecurity across all Member States and the NIS2 Directive will play an important role for achieving this objective.

Disclaimer Note that the authors have assembled the information above to the best of their knowledge as available at the time when the present book was written. Any interpretation is the personal opinion of the authors. Also, the regulation framework can change and evolve any time. The only legally binding text is the one provided in the documents formally published in the European Union Official Journal.

References

1. Warner, M. 2012. "Cybersecurity: A Pre-History." Intelligence and National Security 27 (5): 781–799. https://doi.org/10.1080/02684527.2012.708530

2. Koret, J., & Bachaalany, E. (2015). The antivirus hacker's handbook. John Wiley & Sons.
3. Aycock, J. (2006). Computer viruses and malware (Vol. 22). Springer Science & Business Media.
4. Davies, V. (2021, October 4). The history of cybersecurity. Cyber Magazine. https://cybermagazine.com/cyber-security/history-cybersecurity
5. Furstenau, L., Kremer Sott, M., Homrich, A., Cardoso, T., López-Robles, J., Cobo, M., Abri, A., & Mahlmann Kipper, L. (2020, March). 20 years of scientific evolution of cyber security: a science mapping. In 10th Annual International Conference on Industrial Engineering and Operations Management, https://doi.org/10.46254/AN10.20200376
6. Sihan Qing, Weiping Wen, A survey and trends on Internet worms, Computers & Security, Volume 24, Issue 4, 2005, Pages 334–346, ISSN 0167-4048, https://doi.org/10.1016/j.cose.2004.10.001.
7. Alenezi, M. N., Alabdulrazzaq, H. K., Alshaher, A. A., & Alkharang, M. M. (2022). Evolution of Malware Threats and Techniques: a Review. International Journal of Communication Networks and Information Security (IJCNIS), 12(3). https://doi.org/10.17762/ijcnis.v12i3.4723 (Original work published December 21, 2020)
8. Cristello, Bob. "A Brief History of Cybersecurity," May 23, 2023. https://www.linkedin.com/pulse/brief-history-cybersecurity-robert-cristello/.
9. Willis H. Ware. 1967. Security and privacy in computer systems. In Proceedings of the April 18–20, 1967, spring joint computer conference (AFIPS '67 (Spring)). Association for Computing Machinery, New York, NY, USA, 279–282. https://doi.org/10.1145/1465482.1465523
10. Bell, D. E., & LaPadula, L. J. (1973). Secure computer systems: Mathematical foundations (Vol. 1, p. 2). Technical Report MTR-2547.
11. US Department of Defense. (1985). Department of Defense Trusted Computer System Evaluation Criteria. In: The 'Orange Book' Series. Palgrave Macmillan, London. https://doi.org/10.1007/978-1-349-12020-8_1
12. de Lima, S.M.L., Silva, H.K.d.L., Luz, J.H.d.S. et al. Artificial intelligence-based antivirus in order to detect malware preventively. Prog Artif Intell 10, 1–22 (2021). https://doi.org/10.1007/s13748-020-00220-4
13. Xiao-bin Wang, Guang-yuan Yang, Yi-chao Li and Dan Liu, "Review on the application of artificial intelligence in antivirus detection systemi," 2008 IEEE Conference on Cybernetics and Intelligent Systems, Chengdu, China, 2008, pp. 506–509, https://doi.org/10.1109/ICCIS.2008.4670733
14. European Commission. (2013). Cybersecurity strategy of the European Union: an open, safe and secure Cyberspace (JOIN/2013/01 final). https://eur-lex.europa.eu/legal-content/EN/TXT/HTML/?uri=CELEX:52013JC0001. Accessed 6 Jan 2025
15. Bederna, Z., Rajnai, Z. Analysis of the cybersecurity ecosystem in the European Union. Int. Cybersecur. Law Rev. 3, 35–49 (2022). https://doi.org/10.1365/s43439-022-00048-9
16. "EU Cyber Defence Policy Framework," October 18, 2014. https://ccdcoe.org/uploads/2018/11/EU-141118-EUCyberDefencePolicyFrame-2.pdf
17. Alston & Bird Privacy, Cyber & Data Strategy Blog. "The Digital Single Market Strategy and EU Data Security Policies," July 24, 2017, available at https://www.alstonprivacy.com/the-digital-single-market-strategy-and-eu-data-security-policies/
18. Directive (EU) 2016/1148 of the European Parliament and of the Council of 6 July 2016 concerning measures for a high common level of security of network and information systems across the Union, http://data.europa.eu/eli/dir/2016/1148/oj
19. Directive (EU) 2022/2555 of the European Parliament and of the Council of 14 December 2022 on measures for a high common level of cybersecurity across the Union, amending Regulation (EU) No 910/2014 and Directive (EU) 2018/1972, and repealing Directive (EU) 2016/1148 (NIS2 Directive) [2022] OJ L333/80. https://eur-lex.europa.eu/eli/dir/2022/2555/oj
20. Maria del Mar Negreiro Achiaga. April 20, 2023. "Review of the Directive on security of network and information systems | Legislative Train Schedule," https://www.europarl.europa.eu/legislative-train/theme-a-europe-fit-for-the-digital-age/file-review-of-the-nis-Directive?sid=6901

21. Entwurf eines Gesetzes zur Umsetzung der NIS-2-Richtlinie und zur Regelung wesentlicher Grundzüge des Informationssicherheitsmanagements in der Bundesverwaltung (NIS-2-Umsetzungs- und Cybersicherheitsstärkungsgesetz), 22 July 2024, available at https://www.bmi.bund.de/SharedDocs/gesetzgebungsverfahren/DE/Downloads/kabinettsfassung/CI1/nis2-regierungsentwurf.pdf?__blob=publicationFile&v=1
22. Commission Recommendation of 6 May 2003 concerning the definition of micro, small and medium-sized enterprises (Text with EEA relevance) (notified under document number C(2003) 1422)
23. Directive (EU) 2022/2557 of the European Parliament and of the Council of 14 December 2022 on the resilience of critical entities and repealing Council Directive 2008/114/EC (Text with EEA relevance), available at https://eur-lex.europa.eu/eli/dir/2022/2557/oj
24. Viheriäranta, Henna-Jasmin, Ilona Ilvonen, and Pasi Hellsten. "MANAGING CHANGE IN CYBERSECURITY: NAVIGATING TRANSITION BETWEEN STANDARD VERSIONS AND COMPLIANCE WITH LEGISLATION." Master of Science Thesis, March 2024. https://trepo.tuni.fi/bitstream/handle/10024/155880/ViheriarantaHenna-Jasmin.pdf?sequence=2
25. Wavestone. "NIS 2: Where are the european countries in tranposing the directive? - Wavestone," June 18, 2024. https://wwa.wavestone.com/en/insight/nis-2-european-countries-tranposing-directive/
26. Centre for Cyber Security Belgium. "The NIS2 law" July 22, 2024. https://ccb.belgium.be/en/nis2
27. Zoltán KozmaMárk Almásy, DLA Piper. "NIS2 Directive: The deadline for the registration of entities in Hungary is 30 June 2024 | DLA Piper," n.d. https://www.dlapiper.com/en/insights/publications/2024/06/nis2-directive-the-deadline-for-the-registration-of-entities-in-hungary-is-30-june-2024

Chapter 8
Introduction to European Data Related Regulation

Markus Mueck and Christophe Gaie

Abstract "Data is the new gold"—this is a statement that is often heard in the context of digital services, Artificial Intelligence and the like. There is indeed some truth to it and thus it is essential that data is managed in a fair and transparent manner for all stakeholders. On the other hand, the term "data" is rather abstract and has a multitude of facets. There are so many angles to it that it is challenging to have it covered through a single legislation. In Europe, the management of data is implemented through an interplay of multiple regulations, including the Digital Markets Act, the Regulation on a Single Market for Digital Services, the Data Governance Act, the Data Act, the General Data Protection Regulation (GDPR) and others. Each of those is considering the management of data from a specific perspective. In the present chapter, we give an overview onto selected key regulations in the context of data management. Note that there are further European regulations of relevance to the topic of data and data management but the authors choose to limit the discussion in the present chapter to the above mentioned legal Acts which they consider key.

Keywords Regulation · Data · Digital markets · Governance · Digital services

8.1 Introduction

Data management is a complex task, since on a high level it is an abstract concept spanning a wide area of activities, services, etc. The Data Management Body of Knowledge (DMBOK) effectively illustrates the interconnectedness of different data management dimensions through its data wheel model. At the center of this wheel, data governance plays a crucial role in maintaining balance and ensuring that all data

M. Mueck (✉)
Intel Deutschland GmbH, Munich, Germany
e-mail: Markus.Dominik.Mueck@intel.com

C. Gaie
Head of the Engineering and Digital Innovation Division, Prime Minister Services, Paris, France
e-mail: christophe.gaie@gmail.com

M. Mueck and C. Gaie (eds.), *European Digital Regulations*, Intelligent Systems Reference Library 265, https://doi.org/10.1007/978-3-031-80809-8_8

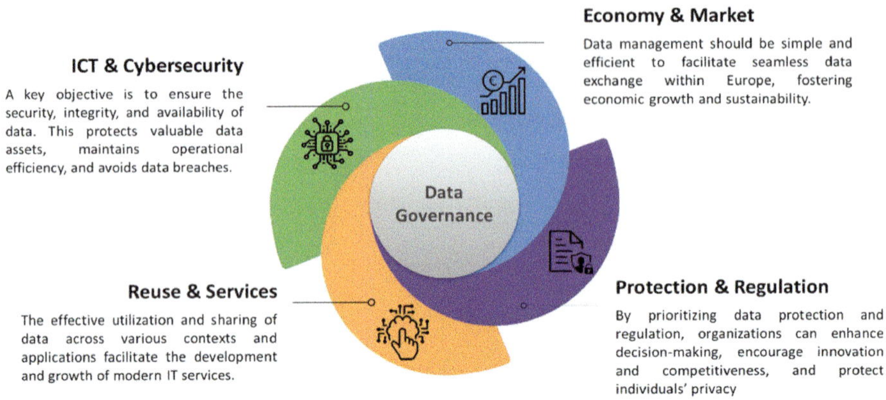

Fig. 8.1 Data management dimensions to consider in the context of European data regulation

activities align with organizational objectives [1]. Inspired by this model, Fig. 8.1 illustrates the data management dimensions that apply for European Data regulation.

Data governance encompasses several key dimensions:

- **Economy and Market:** Data management should be streamlined and efficient to facilitate seamless data exchange within Europe, fostering economic growth, and sustainability. This includes investing in modern data infrastructure, promoting interoperability standards, ensuring property rights and supporting the development of data-driven industries [2].
- **Protection and Regulation:** Prioritizing data protection and regulation ensures ethical data use, mitigates risks, and builds trust, ultimately enhancing decision-making, encouraging innovation, and protecting individuals' privacy [3, 4]. This involves implementing robust security measures, complying with relevant laws and regulations, and fostering a culture of data ethics.
- **ICT and Cybersecurity:** It is necessary to ensure the security, integrity, and availability of data. This protects valuable data assets, maintains operational efficiency, and prevents data breaches. Investing in cybersecurity technologies, training personnel, and developing incident response plans are essential components of this dimension.
- **Reuse and Services:** The effective utilization and sharing of data across various contexts and applications are crucial for the development and growth of contemporary IT services [5]. This involves developing data sharing platforms, promoting data interoperability, and fostering a culture of data-driven innovation.

By addressing these dimensions comprehensively, organizations can establish a robust data governance framework that supports their strategic objectives, creates value, and contributes to a thriving digital economy.

In Europe, a number of key Regulations are introduced addressing requirements on data management from different angles. Obviously, there are interdependencies between those Regulations.

In the context of the present chapter, the authors choose to focus on the Digital Markets Act [6], the Digital Services Act [7], the Data Governance Act [8], the Data Act and the General Data Protection Regulation (GDPR) [9]. While these regulations are tackled separately in the literature [10–12] there is, to the authors' knowledge, no global and comprehensive approach of the topic.

It is understood that the topic may be further addressed in other Regulations to a more or less explicit extent. Still, those Regulations are establishing key pillars on data management in Europe. For example, the Artificial Intelligence Act (AI Act) may also be considered an important regulation addressing Data Management from the specific angle of Artificial Intelligence systems as illustrated in Fig. 8.2.

In the following sections of the present chapter, a summary is given on the key Regulations related to Data Management as listed above.

The construction of this regulatory framework was progressively built over the past decades. First, pioneering countries defined national Data Protection and Privacy Regulations: Sweden 1973 [13], Germany 1977 [14]. In 1995, the EU introduced Directive 95/46/EC, the first major law to protect personal data [15]. It laid

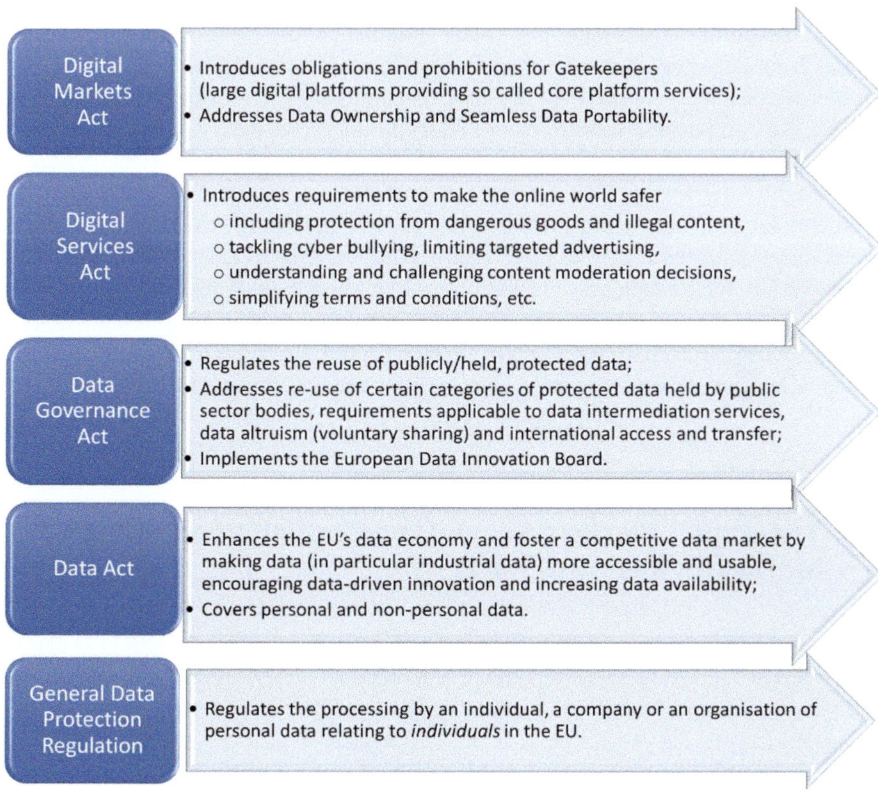

Fig. 8.2 An overview of European key regulations (selection) related to data management

down essential rights for individuals, like the ability to access and correct their personal information, and emphasized that data should be handled fairly and legally. Fast forward to 2016, the General Data Protection Regulation (GDPR) came into play, replacing the earlier directive. This new regulation was more robust, giving people greater control over their personal data and placing tougher obligations on organizations that handle it [16].

In 2022, the Digital Services Act (DSA) was introduced to regulate online platforms and marketplaces within the EU [17]. Its goals include protecting users by requiring platforms to tackle illegal content and curb the spread of misinformation and harmful material. The DSA also enhances transparency, as online platforms must be clearer about their content moderation practices and how they use algorithms. Additionally, large platforms with over 45 million users in the EU need to evaluate and address risks related to illegal content and manipulative practices on their services. The DSA works in conjunction with the GDPR to foster a safer and fairer online environment.

Alongside the DSA, the Digital Markets Act (DMA) was also adopted in 2022 [18]. This regulation aims to ensure contestable and fair markets in the digital sector, addressing the power dynamics between large digital platforms, known as "gatekeepers," and other businesses. The DMA establishes rules that prevent these gatekeepers from engaging in anti-competitive practices, promoting fairness in how they operate. It complements the DSA by focusing not only on user protection but also on fostering competitive conditions in the digital marketplace.

Lastly, the Data Governance Act, adopted in 2022, focuses on how data is shared and reused across Europe [19]. It aims to promote a common data space and sets guidelines for data intermediaries and altruistic data sharing. It was followed, in 2023, by the adoption of the Data Act [20]. This regulation aims to ensure that individuals and businesses have fair access to and control over their data. It also seeks to level the playing field among businesses by promoting equitable data sharing. An interesting historic approach that also tackles non-European countries (United States of America, United Kingdom, Asian countries, …) was proposed by Lonzetta and Hayajneh [21] (Fig. 8.3).

8.2 The Digital Markets Act and the Regulation on a Single Market for Digital Services

The Digital Markets Act (DMA) [6] and the Digital Services Act (DSA) [7] can be understood as a package with the objective *to create a safer digital space where the fundamental rights of users are protected and to establish a level playing field for businesses* [22].

Indeed, the two Acts *create a safer digital space in which the fundamental rights of all users of digital services are protected and establish a level playing field to*

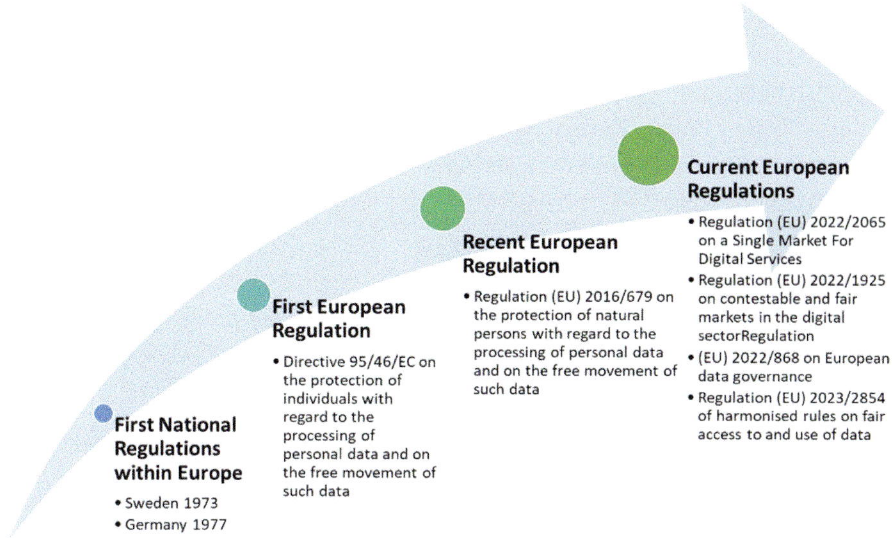

Fig. 8.3 A timeline of European data legislations

foster innovation, growth, and competitiveness, both in the European Single Market and globally [22].

Differentiated targets of the Digital Markets Act (DMA) [6] and the Digital Services Act (DSA) [7] are illustrated in Fig. 8.4.

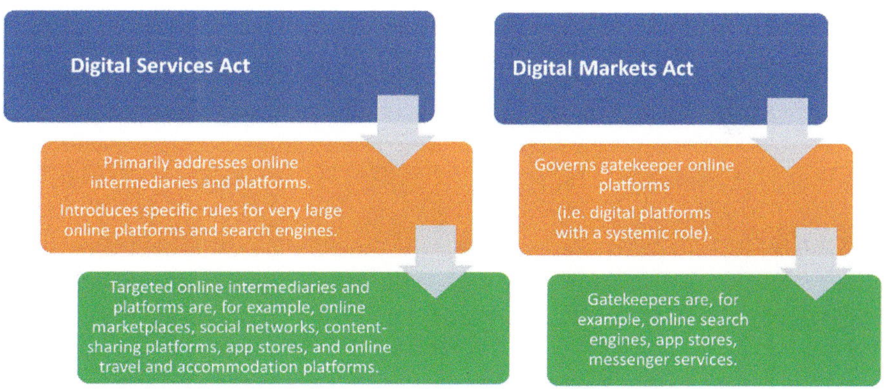

Fig. 8.4 High level overview on the digital services act [7] and the digital markets act [6]

8.2.1 The Digital Markets Act (DMA)

The Digital Markets Act (DMA) [6] introduces the concept of "Gatekeepers" and establishes related objective criteria. Gatekeepers are large digital platforms providing so called core platform services, such as for example online search engines, app stores and messenger services. Gatekeepers will have to comply with the obligations and prohibitions listed in the DMA [23].

Specifically, the Digital Markets Act (DMA) defines 'Gatekeeper' as undertaking one of the core platform services displayed in Fig. 8.5.

CHAPTER III (practices of Gatekeepers that limit contestability or are unfair), Article 5 (Obligations for gatekeepers) of the Digital Markets Act (DMA) [6] defines the Do's and Don'ts for Gatekeepers. A selection of key requirements is illustrated in Fig. 8.6.

The upper rules for Gatekeepers are leading to a number of benefits for Users [24], some of which are summarized in Fig. 8.7.

From a timeline perspective, the initial proposal of the European Commission was published in December 2020, the Regulation was adopted by the European

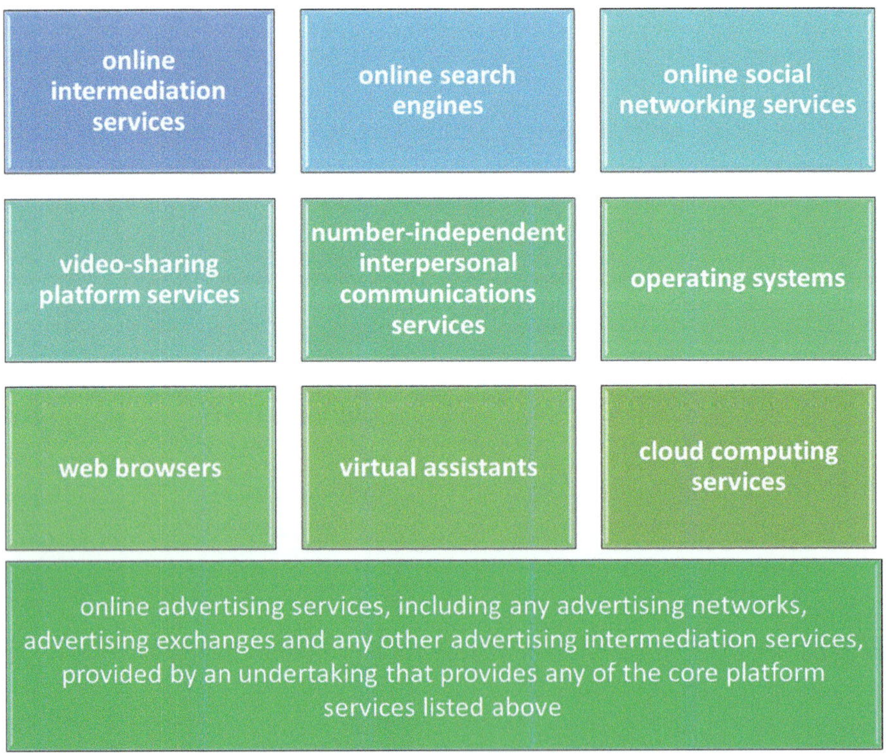

Fig. 8.5 Gatekeepers undertaking one of the above outlined core platform services as defined by the digital markets act [6]

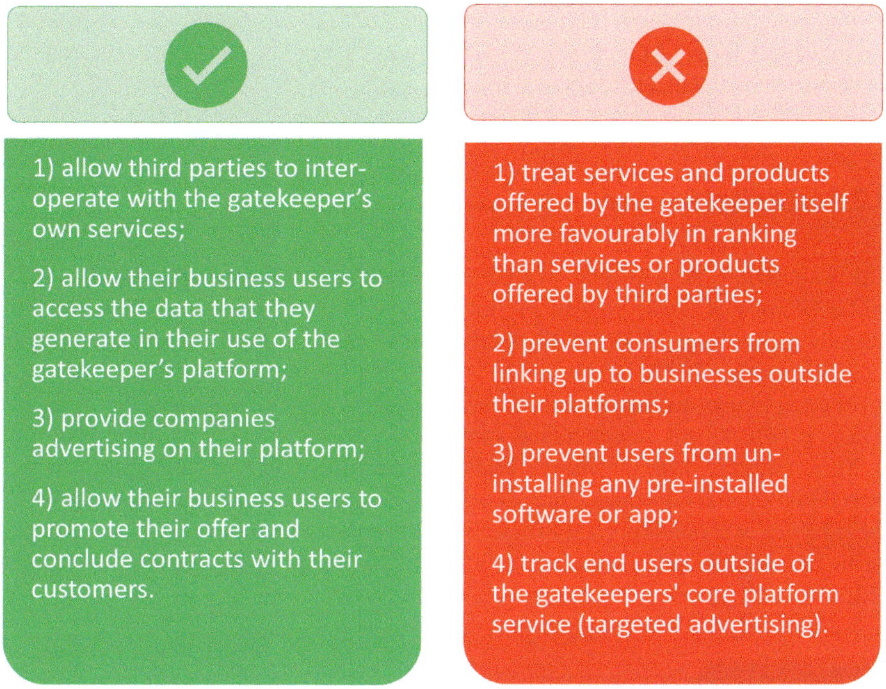

Fig. 8.6 Do's and don'ts (selection) for Gatekeepers as defined by the digital markets act [6, 24]

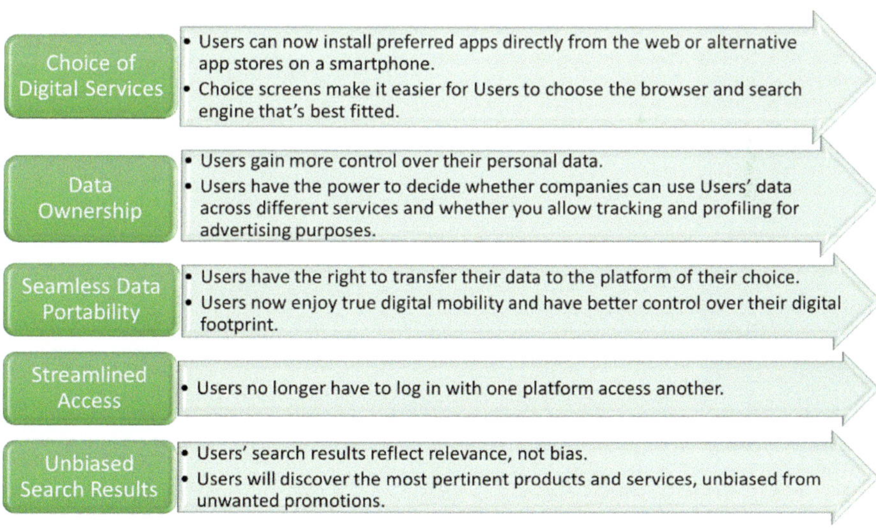

Fig. 8.7 Summary of advantages for users from the digital markets act [6, 24]

Parliament and the Council on 14 September 2022. It was finally published in the Official Journal on 12 October 2022. The Digital Markets Act (DMA) entered into force on 1 November 2022 and has become applicable on 2 May 2023 [24].

8.2.2 *Digital Services Act (DSA)*

The Regulation on a Single Market for Digital Services, i.e. Regulation (EU) 2022/2065 [7] of the European Parliament and of the Council of 19/10/2022 and amending Directive 2000/31/EC (Digital Services Act—DSA), entered into force on 16 November 2022.

In complement to the Digital Markets Act (DMA) [6], the Digital Services Act (DSA) [7] introduces safeguards as summarized in Fig. 8.8 [25].

To summarize, with the Digital Markets Act (DMA) [6] and the Digital Services Act (DSA) [7], the European Union is creating a modern legal framework that protects the safety of users online, establishes governance with the protection of fundamental rights at its forefront, and maintains a fair and open online platform environment [22].

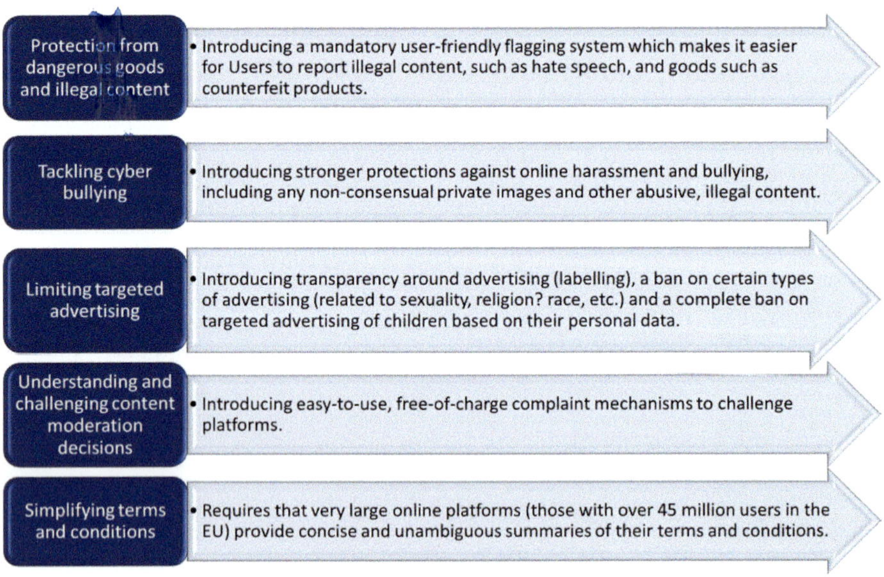

Fig. 8.8 Summary of objectives of the digital services act [7, 25]

8.3 The Data Governance Act

The Data Governance Act [8], which became effective on June 23, 2022, and has been applicable since September 2023, aims to boost confidence in data sharing, enhance mechanisms for increasing data availability, and address technical hurdles to data reuse [26].

Hereby, the Data Governance Act [8] supports the concept of Common European Data Spaces in strategic domains, involving both private and public players, across various sectors including health, environment, energy, agriculture, mobility, finance, manufacturing, public administration, and skills [26]. The Data Governance Act [8] complements the Open Data Directive [27], regulating the reuse of publicly available information held by the public sector. Notably, there is protected data held by the public sector that cannot be made openly available.

Article 1 of the Data Governance Act [8] defines its main scope as displayed in Fig. 8.9.

CHAPTER II (Re-use of certain categories of protected data held by public sector bodies), Article 3 (Categories of data) defines the application of data held by public sector bodies which are protected on grounds illustrated in Fig. 8.10.

The expectation is that the Data Governance Act will increase data availability and facilitate data sharing among sectors and EU countries in order to maximize the potential of data for the benefit of European citizens and businesses [26].

In order to achieve those objectives, the European Union will take the four broad set of measures [26] highlighted in Fig. 8.11.

Data intermediation services are introduced to address the concern of stakeholders, especially private companies, that sharing their data would *imply a loss of competitive advantage and represent a risk of misuse* [28]. The Data Governance Act [6] introduces data intermediation services (also called data intermediaries, including

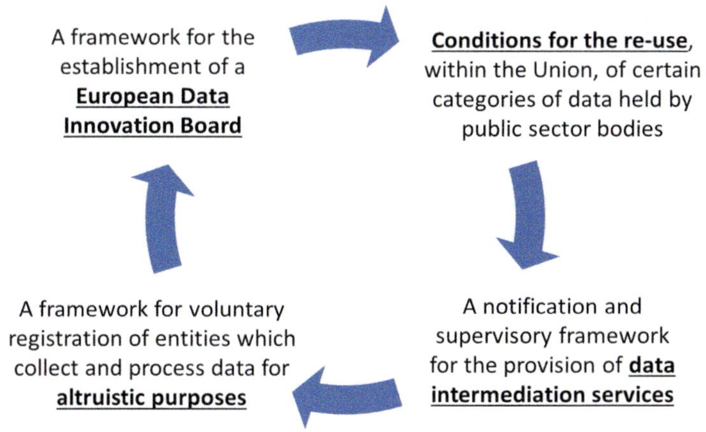

Fig. 8.9 Scope (selection) of the data governance act [8]

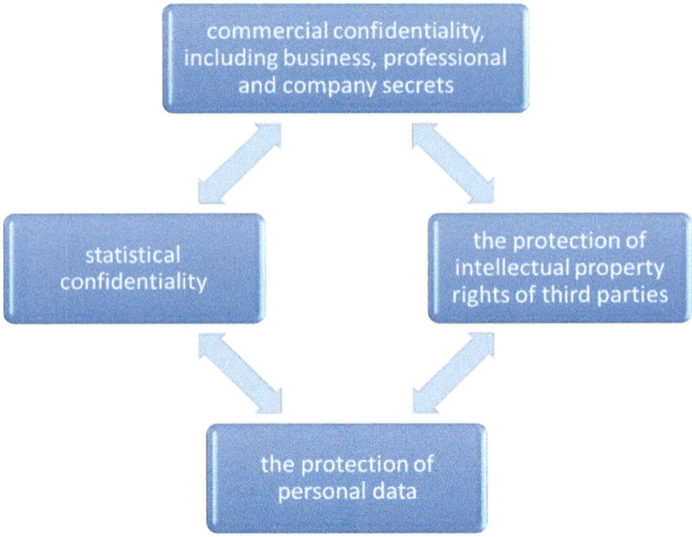

Fig. 8.10 Categories of data as defined by the data governance act [8]

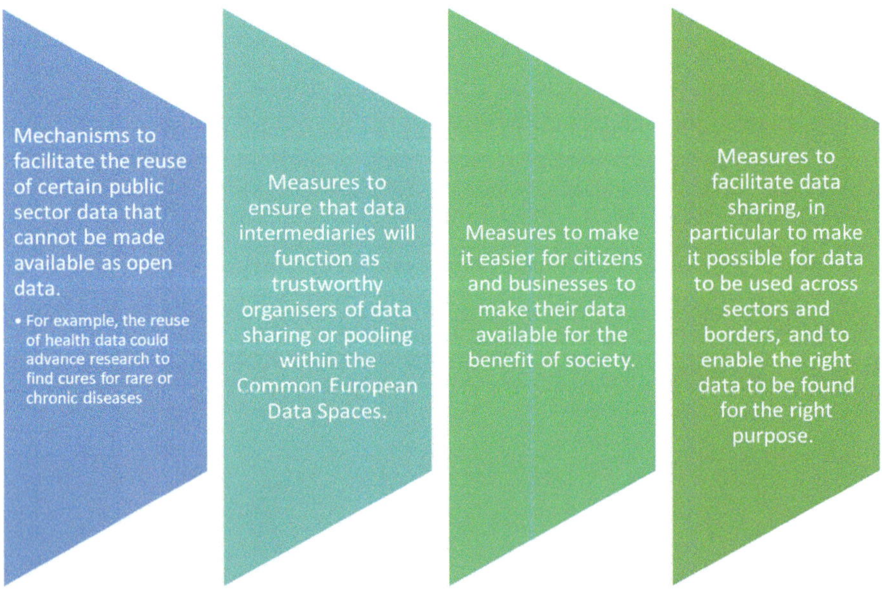

Fig. 8.11 Set of measures by European Union in context of the data governance act [8, 26]

data marketplaces) based on a model guaranteeing neutrality and transparency, while maintaining stakeholders in control of their data.

A provider of data intermediation services will act as neutral third party connecting stakeholders with data users. Such providers can charge for facilitating the data sharing, but they are not authorized to use the managed data for financial profit, such as selling data, etc. Providers of data intermediation services are required to register at the competent authorities and are closely supervised by those, ensuring that the service is non-discriminatory, does not distort competition, etc. A central EU register of data intermediation services is established as explained in [29].

Data altruism is defined by the Governance Data Act [6]. To summarize, the concept's key idea is about the voluntary sharing of data allowing the use of stakeholders' non-personal data without obtaining a reward for objectives of general interest, including strengthening healthcare services, addressing climate change, mobility, facilitating the handling of official statistics, improving public services, public policy making or scientific research of general interest. For this purpose, the Governance Data Act [6] introduces data altruism organisations (recognised in the EU) which must be of not-for-profit character and meet transparency requirements, among other requirements. They are contained in the EU register of recognised data altruism organisations as explained in [30].

International data flows are acknowledged as a key concept by the European strategy for data of February 2020 [31]. The Data Governance Act complements the General Data Protection Regulation (GDPR) [9] which specifically introduces safeguards for the protection of personal data. Especially, attention is given to access requests from 3rd country governments concerning non-personal data, including public sector data, data intermediation services and data altruism constellations. Also, the European Commission is empowered by the Data Governance Act [6] to make available model contract clauses for public sector bodies and re-users for scenarios where public sector data is involved in data transfers with third countries [28].

The European Data Innovation Board was established by the European Commission as provided by the Governance Data Act [6]. It supports the sharing of best practices, with specific attention to data intermediation, data altruism, and the use of public data that cannot be made available as open data, as well as on the prioritization of cross-sectoral interoperability standards [28]. Specific power is given to the European Data Innovation Board which will propose guidelines for Common European Data Spaces, for example on the adequate protection for data transfers outside of the EU [28].

8.4 The Data Act

The EU Data Act [32], i.e. Regulation (EU) 2023/2854, was published in the Official Journal of the European Union on 22 December 2023. Following its entry into force, the Data Act will become applicable after 20 months on 12 September 2025.

Fig. 8.12 Scope of the data act [32]

The Data Act [32] was created to support the EU's data economy and drive a competitive data market by making industrial and other types of data more accessible and usable, growing data-driven innovation, and enhancing data availability. As stated by the European Commission [33], the Data Act ensures fairness in the allocation of the value of data among the actors in the data economy. It clarifies who can use what data and under which conditions. Users of connected products (businesses or individuals that own, lease, or rent such a product) are given greater control over the data they generate, while maintaining incentives for those who invest in data technologies. In addition, general conditions are introduced for situations where a business has a legal obligation to share data with another business.

As illustrated in Fig. 8.12, Article 1 of the Data Act [32] is stating that scope of the Regulation, which is to lay down *harmonised rules, inter alia, on*:

Furthermore, the Data Act [32] covers personal and non-personal data. It is concerning manufacturers of connected products to be placed on the market in the EU and providers of related services, users in the EU of connected products or related services, data holders that make data available in the EU (irrespective of their place of establishment), EU data recipients, public sector bodies requiring data holders to make data available, providers of data processing services (irrespective of their place of establishment), providing such services in the EU and participants in data spaces and vendors of applications using smart contracts and persons whose trade, business or profession involves the deployment of smart contracts.

The Data Act [32] is structured into eight main chapters which are summarized in Figs. 8.13 and 8.14.

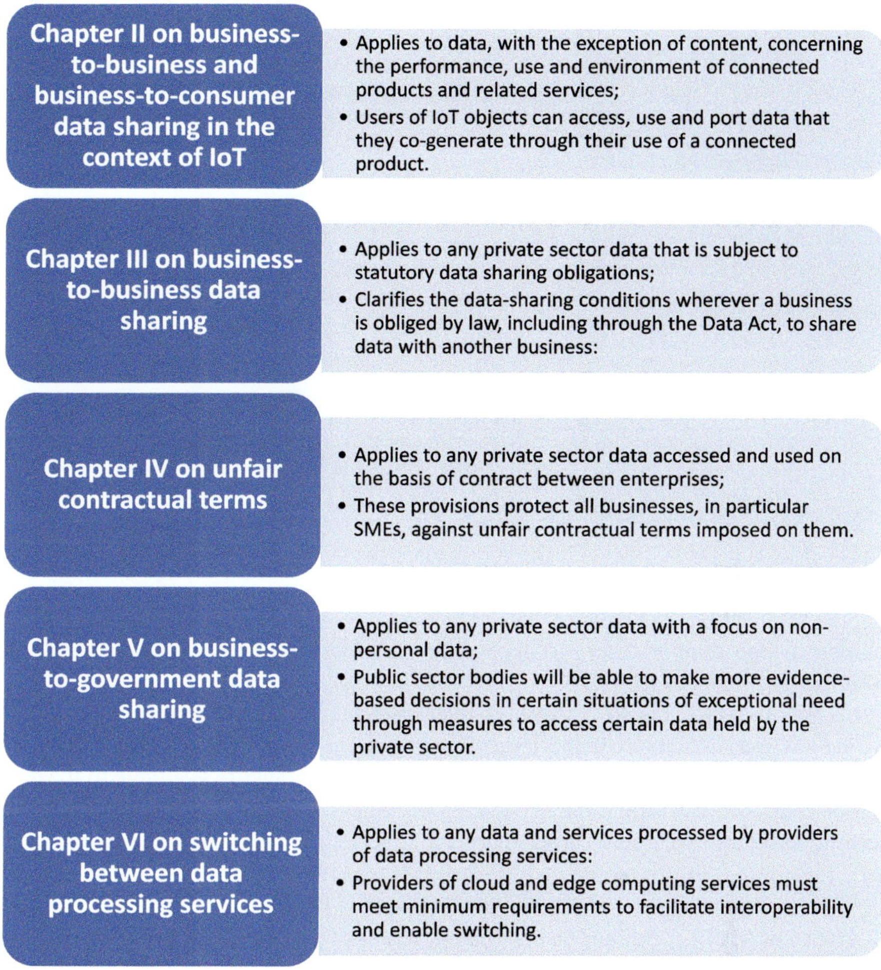

Fig. 8.13 Overview on data act [32], Chapters II–VI

8.4.1 Chapter II: Business-to-Business and Business-to-Consumer Data Sharing in the Context of the IoT Market

As a priority of the legislation, the Data Act [32] targets the promotion of fairness in the data economy and empower users to benefit from the data they generate through connected products they own, rent, or lease [33]. The Act allows users of connected devices—such as cars, medical and fitness devices, and industrial or agricultural machinery—as well as related services (like apps that adjust light brightness or

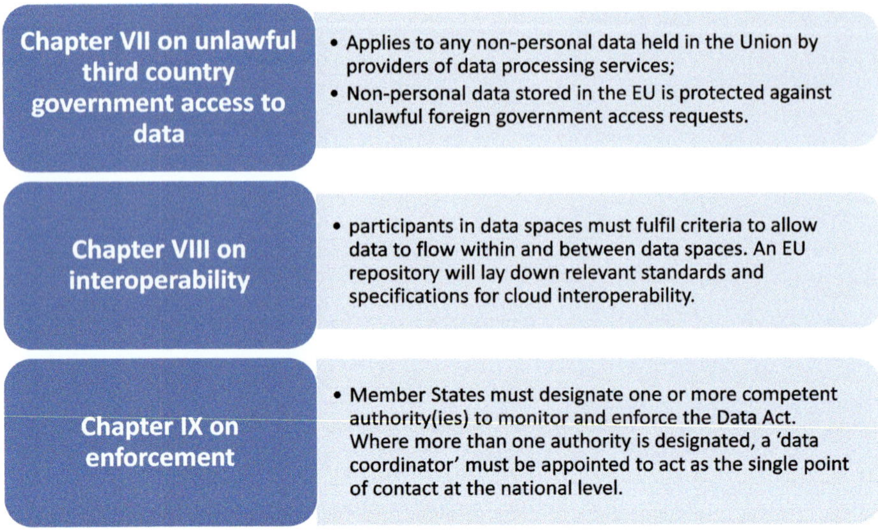

Fig. 8.14 Overview on data act [32], Chapters VII–IX

regulate fridge temperatures) to access the data they co-create while using these products and services [33].

Chapter II of the Data Act focuses on data sharing between businesses and between businesses and consumers. It applies to all raw and pre-processed data generated from the use of connected products or related services that are readily available to the data holder (e.g., the manufacturer or service provider). This means data that can be accessed without excessive effort. The Act covers both personal and non-personal data, including relevant metadata [33], but excludes inferred or derived data, such as highly enriched analytics or multimedia content.

The Data Act [32] enables Users to have access to data that they generate through the usage of a connected product or related services. In case that the Users wish to share their data with another stakeholder, they can do so either directly or by instructing the holder of the data to execute the sharing. The data holder (typically the manufacturer of a connected product or a provider of a related service) is obliged to have a contract with the User defining access, use and sharing rights of the data. The data holder must seek the agreement of the User to use any non-personal data generated by the User. Data holders must make the data available to users for free.

The basic principles are illustrated in Fig. 8.15.

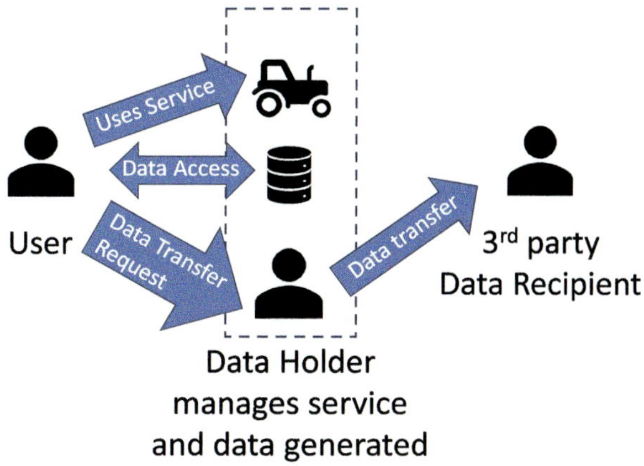

Fig. 8.15 Data act [32] enabling access to data generated by user through the usage of a connected product or related services

8.4.2 *Chapter III: Rules on Mandatory Business-to-Business Data Sharing*

Under the Data Act [32], there may be situations where a Data Holder has the obligation to make data available to another stakeholder (Data Recipient), for example in the context of IoT data. As a general principle, fairness must be ensured for data sharing terms and conditions, reasonable and non-discriminatory. It is possible, tough, for Data Holders to request an appropriate compensation in case that they obliged to transfer data to a Data Recipient.

All rules under Chapter III of the Data Act apply to personal and non-personal Data.

8.4.3 *Chapter IV: Unfair Contractual Terms*

In the context of Data management, there may be situations where (smaller) businesses, such as SMEs, need to negotiate access to data with other stakeholders which are in a stronger bargaining position. While contractual freedom is understood to be central to business-to-business relationships, the Data Act [32] introduces safeguards to protect against unfair contractual terms, such as for example non-negotiable terms ('take-it-or-leave-it') related to data access and data use.

Both personal and non-personal data are covered by these rules, held by a private entity that is accessed and used based on a contract between businesses.

8.4.4 Chapter V: Business-to-Government Data Sharing

In exceptional circumstances, government organizations may be authorized to access data held by private stakeholders to fulfill a task of public interest. Such exceptional needs can arise during public emergencies—like major natural disasters, pandemics, or cybersecurity incidents—as well as in non-emergency situations. For example, aggregated and anonymized data from drivers' GPS systems could be utilized to optimize traffic flows [33].

Under Chapter V, all data are in scope, with a focus on non-personal data.

8.4.5 Chapter VI: Switching Between Data Processing Services

With the objective to create a competitive European Market, User of Data Processing Services, explicitly including Cloud and Edge services, the Data Act [32] foresees the possibility to let Users switch seamlessly between providers. As of today, in practice, Users face barriers when they seek to change providers, such as high cost, complex transfer procedures, etc. Under the Data Act, the switching will be free, fast and fluid. Note that the Data Act will *entirely remove switching charges, including charges for data egress (i.e. charges for data transit), from 12 January 2027* [33].

Chapter VI of the Data Act applies to providers of data processing services.

8.4.6 Chapter VII: Unlawful Third Country Government Access

We consider the case that a Government outside of the European Union has access granted to and is able to transfer of non-personal data processed and stored within the EU. The Data Act [32] introduces safeguards making sure that related access is only granted in case that the request does not *conflict with EU laws and guarantees on the protection of the fundamental rights of individuals, national security interests or commercially sensitive data* [33]. In this context, the Data Act [32] follows the principles of the Data Governance Act [8], *preventing unlawful third-country governmental access and transfer of non-personal data held in the EU* [33].

Chapter VII applies to any non-personal data held in the EU by a provider of a data processing service.

8.4.7 Chapter VIII: Interoperability

The Data Act [32] requires participants in data spaces to comply with a number of essential requirements, such that a standardized exchange of data is available in Common European Data Spaces. Such a standardized exchange may rely on data structures, data formats and vocabularies defined in harmonised standards and open interoperability specifications. Further details may be implemented by the European Commission through Delegated Acts.

Chapter VIII addresses participants of data spaces offering data or data-based services to other participants, facilitating or engaging in data sharing within the data spaces. Also, vendors of smart contracts and data processing services providers are addressed.

8.4.8 Chapter IX: Enforcement and Overarching Provisions and the European Data Innovation Board (EDIB)

The Data Act [32] requires Member States to designate a "data coordinator", i.e. a competent authority to ensure the efficient implementation of the Data Act. Furthermore, Member States have the option to establish *certified dispute settlement bodies to assist parties who cannot agree on fair, reasonable and non-discriminatory terms for making data available* [33].

Article 42 of the Data Act [32] refers to the European Data Innovation Board (EDIB) established by the Commission as an expert group pursuant to Article 29 of the Data Governance Act [8], i.e. Regulation (EU) 2022/868. The European Data Innovation Board is required to support the consistent application of this Regulation by measures given in the Article.

This includes discussions among competent authorities to coordinate and establish recommendations for setting penalties for infringements of the Data Act. Penalties are determined by these authorities, and the Data Act stipulates that they should be effective, proportionate, and dissuasive [33].

8.4.9 Essential Requirements of the EU Data Act

Under CHAPTER VIII: INTEROPERABILITY, the EU Data Act introduces essential requirements in Article 33, Article 35 and Article 36 as illustrated in Fig. 8.16.

Article 33 addresses *essential requirements regarding interoperability of data, of data sharing mechanisms and services, as well as of common European data spaces.* The European High Level Forum (HLF) has issued a related report entitled *HLF Workstream 14 on Data Interoperability—Final Report* [34].

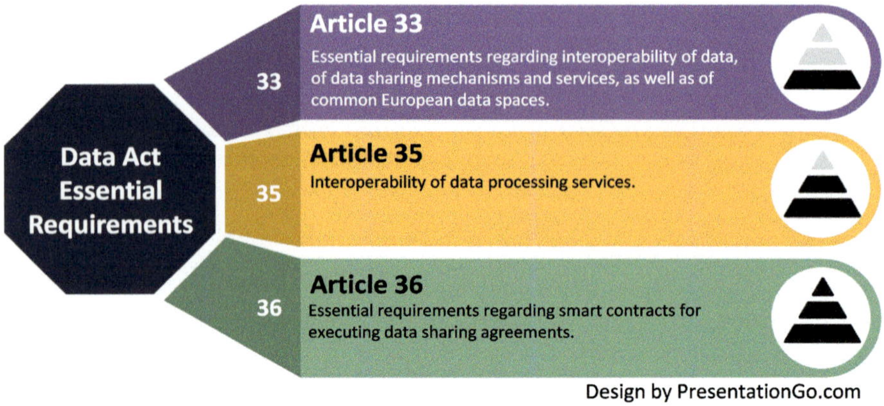

Fig. 8.16 Data act [32] articles introducing essential requirements

Below, key parts of Article 33(1) are summarized which introduced essential requirements which apply to *participants in data spaces*. It is indeed noteworthy that those essential requirements apply to the *participants* and not to the *data spaces themselves*.

CHAPTER VIII

Interoperability

Article 33

Essential requirements regarding interoperability of data, of data sharing mechanisms and services, as well as of common European data spaces

1. **Participants in data spaces** that offer data or data services to other participants shall comply with the **following essential requirements** to facilitate the interoperability of data, of data sharing mechanisms and services, as well as of common European data spaces ...:
   (a) the dataset content, use restrictions, licences, data collection methodology, data quality and uncertainty shall be sufficiently described, where applicable, in a machine-readable format, to allow the recipient to **find, access and use the data**;
   (b) the **data structures, data formats, vocabularies**, classification schemes, taxonomies and code lists, where available, shall be described in a publicly available and consistent manner;
   (c) the technical means to **access the data**, such as application programming interfaces, and their terms of use and quality of service shall be sufficiently described to enable automatic access and transmission of data between parties, ...;

(d) where applicable, the means to enable the interoperability of tools for automating the execution of **data sharing agreements**, such as smart contracts shall be provided.

The requirements can have a generic nature or concern specific sectors, ...

Article 35 addresses *interoperability of data processing services*. Below, key parts of Article 35(1) and Article 35(2) are summarized:

CHAPTER VIII

Interoperability

Article 35

Interoperability of data processing services

1. **Open interoperability specifications and harmonised standards** for the interoperability of data processing services shall:

   (a) achieve, where technically feasible, **interoperability between different data processing services** that cover the same service type;
   (b) enhance **portability of digital assets** between different data processing services that cover the same service type;
   (c) facilitate, where technically feasible, **functional equivalence between different data processing services** referred to in Article 30(1) that cover the same service type;
   (d) not have an adverse impact on the security and integrity of data processing services and data;
   (e) be designed in such a way so as to allow for technical advances and the inclusion of new functions and innovation in data processing services.

2. Open interoperability specifications and harmonised standards for the interoperability of data processing services shall adequately address:

   (a) the **cloud interoperability aspects** of transport interoperability, syntactic interoperability, semantic data interoperability, behavioural interoperability and policy interoperability;
   (b) the **cloud data portability aspects** of data syntactic portability, data semantic portability and data policy portability;
   (c) the **cloud application aspects** of application syntactic portability, application instruction portability, application metadata portability, application behaviour portability and application policy portability.

...

Article 36 addresses *essential requirements regarding smart contracts for executing data sharing agreements*. Below, key parts of Article 36(1) are summarized:

CHAPTER VIII

Interoperability

Article 36

Essential requirements regarding smart contracts for executing data sharing agreements

1. The **vendor of an application using smart contracts** ... to make data available shall ensure that those smart contracts comply with the following **essential requirements** of:
   (a) **robustness and access control**, to ensure that the smart contract has been designed to offer access control mechanisms and a very high degree of robustness to avoid functional errors and to withstand manipulation by third parties;
   (b) **safe termination and interruption**, to ensure that a mechanism exists to terminate the continued execution of transactions ...;
   (c) **data archiving and continuity**, to ensure, in circumstances in which a smart contract must be terminated or deactivated, there is a possibility to archive the transactional data, smart contract logic and code ...;
   (d) **access control**, to ensure that a smart contract is protected through rigorous access control mechanisms at the governance and smart contract layers; and
   (e) **consistency**, to ensure consistency with the terms of the data sharing agreement that the smart contract executes.

....

A corresponding Standardisation Request is expected to be issued in the future to European Standardisation Organisations (ESOs) to develop Harmonised European Norms—or Harmonised Standards—providing presumption of conformity.

8.5 General Data Protection Regulation (GDPR)

The European Parliament and Council of the European Union adopted the General Data Protection Regulation (GDPR) [9] on 14 April 2016, it entered into force on 24 May 2016 and applies since 25 May 2018. It replaces the 1995 Data Protection Directive [35] which was adopted at a time when the internet was in its infancy. The

GDPR [9] has substantial influence on a global level, having for example profoundly impacted The California Consumer Privacy Act (CCPA) [36].

The General Data Protection Regulation (GDPR) [9] governs the processing of personal data by individuals, companies, or organizations concerning individuals in the EU. It does not cover the processing of personal data related to deceased persons or legal entities. Additionally, the rules do not apply to data processed by individuals for purely personal purposes or activities conducted in the home, as long as there is no connection to professional or commercial activities. However, if an individual uses personal data outside the personal sphere, such as for socio-cultural or financial activities, then data protection laws must be observed [37].

Based on the European Commission information site [38], the illustration of Fig. 8.17 summarizes a selection of the rights for citizens under the General Data Protection Regulation (GDPR) [9]:

Based on the European Commission information site [38], a summary of a selection of rules for business and organizations is outlined in Figs. 8.18 and 8.19.

8.6 Conclusion

Data related Regulation is a complex and multi-faceted task. Key European regulations in this field are the Digital Markets Act [6], the Digital Services Act [7], the Data Governance Act [8], the Data Act, the General Data Protection Regulation (GDPR) [9] and others. Europe has indeed created a comprehensive framework to protect data ownership and related data management while allowing for an efficient operation of the European Union Single Market. The General Data Protection Regulation (GDPR), in particular, has major impact around the globe and is substantially influencing related debates in other regions. The Data Act is expected to follow these foot-steps by enhancing the EU's data economy and fostering a competitive data market.

Further directions of data regulations should consider the interactions with United States and to facilitate international cooperation. Another key aspect to consider is the capability of stakeholders to comply to the evolution of the regulations. A valuable approach is to ensure backward compatibility as far as possible. This contributes to the competitiveness of the economy as well as a continuous improvement of stakeholders for a better usage and protection of data.

Disclaimer Note that the authors have assembled the information above to the best of their knowledge as available at the time when the present book was written. Any interpretation is the personal opinion of the authors. Also, the legislation framework can change and evolve any time. The only legally binding text is the one provided in the documents formally published in the European Union Official Journal.

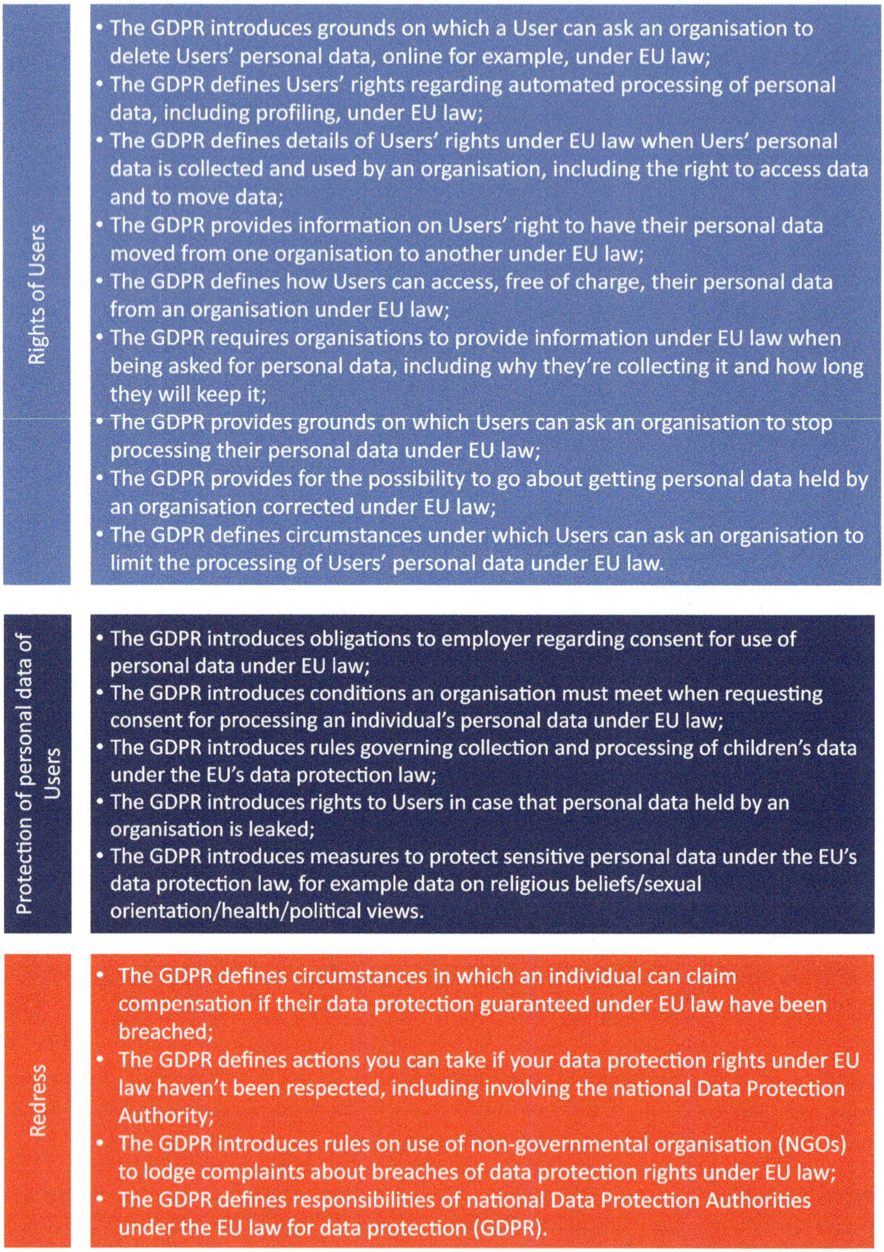

Fig. 8.17 Summary of a selection of the rights for citizens [38] under the General Data Protection Regulation (GDPR) [9] (shortened)

Application of the regulation	• The GDPR introduces rules which only apply to personal data about individuals, they don't govern data about companies or any other legal entities; • The GDPR defines circumstances in which the EU data protection rules apply to SMEs; • The GDPR defines which organisations, including SMEs , have to respect the EU's data protection law. Specifically, the GDPR applies to i) a company or entity which processes personal data as part of the activities of one of its branches established in the EU, regardless of where the data is processed; or ii) a company established outside the EU and is offering goods/services (paid or for free) or is monitoring the behaviour of individuals in the EU.
Dealing with citizens	• The GDPR introduces grounds on which automated decision-making is permitted or not under EU data protection law; • The GDPR introduces rules, including deadlines, for responding for citizens' request made under the EU's data protection law; • The GDPR introduces grounds on which people can ask for their data to be moved from your organisation to another organisation under the EU data protection law; • The GDPR defines what to do if a citizen objects to you processing their data, for, for example, marketing purposes, under the EU data protection law; • The GDPR defines grounds on which an organisation can refuse, or not, to delete a citizen's personal data under the EU's data protection law; • The GDPR defines the type of data citizens can access on request and what they have to receive from your organisation under the EU data protection law.
Enforcement and sanctions	• The GDPR establishes the European Data Protection Board (EDPB) that works to help ensure that the data protection law is applied consistently across the EU; • The GDPR defines the role and responsibilities of national Data Protection Authorities in enforcing and monitoring the EU data protection law; • The GDPR introduces an organisation's liability for individual damages in case of non-compliance with the EU's data protection law. A range of sanctions, including suspension of activities and fines can be imposed if a company doesn't comply with EU law on data protection.
Legal grounds for processing data	• The GDPR defines that children's personal data can only be collected and processed on the grounds of consent with the consent of a parent or guardian under EU law; • The GDPR introduces rules governing use of people's personal data for direct marketing under the EU's data protection law, the GDPR; • Data protection rules post May 2018 change the rules on consent, consent given pre-May 2018 is not always valid; • The GDPR requires that a person can withdraw consent for the processing of their data under EU law, in certain circumstances. Organisations should enable them to do this; • The GDPR requires that consent for processing personal data for scientific research is subject to specific rules under EU data protection law; • EU data protection rules set down conditions as to when an organisation can process an individual's data, including with consent or a contract;
Obligations	• Obligations for organisations under EU data protection law are based on the type and volume of their activity not their size; • The GDPR defines Tools available to check if an organisation adheres to the EU's data protection law; • The GDPR defines what rules apply if an organisation transfers data outside the EU; • Under the EU's data protection law data protection has to be built into the early stages of product design (data protection 'by design' and 'by default'); • The GDPR defines when what companies have to carry out data impact assessments and how; • The GDPR introduces rules on what to do in case of a data breach; • The GDPR defines circumstances in which another company can process data on your behalf under the EU's data protection law and precautions to take; • The GDPR defines how the data controller and data processor is determined and the responsibilities of each under the EU data protection regulation.
Principles of the GDPR	• The GDPR defines rules on the length of time personal data can be stored and whether it needs to be updated under the EU's data protection rules; • The GDPR requires that data cannot be processed for any purpose. The purpose for processing of personal data must be known and the individuals whose data you're processing must be informed; • The GDPR defines conditions when data can be used for a different purpose to that for which it was collected; • The GDPR introduces rules on volumes of data that can be collected from individuals under the EU data protection law. • The GDPR requires that a list of the type of information organisations must provide citizens with when collecting their data, this includes who is collecting it and why.
Public administrations and data protection	• The GDPR requires that individuals have rights under EU data protection law, including the right to access data, public administrations have to deal with these requests; • The GDPR requires that public administrations have to respect the key principles of EU data protection law, including informing individuals. • The GDPR introduces sanctions if a public administration doesn't respect EU data protection law may include suspension of activity, warnings or fines.

Fig. 8.18 Summary (part 1) of a selection of rules for business and organizations [38] under the General Data Protection Regulation (GDPR) [9] (shortened)

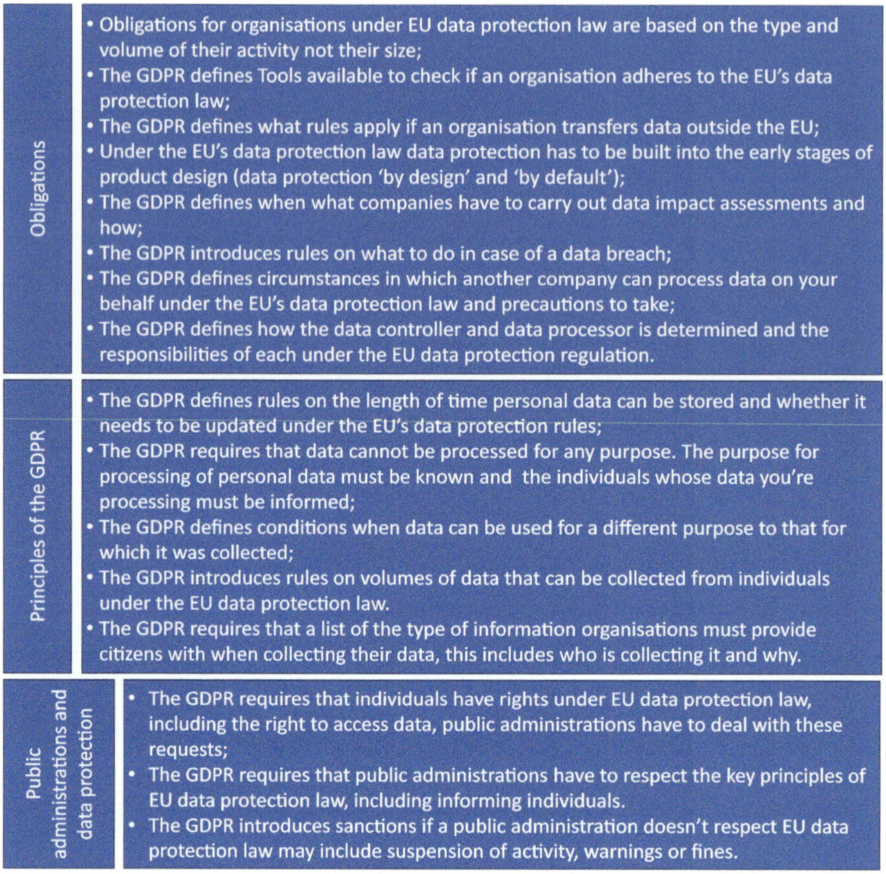

Fig. 8.19 Summary (part 2) of a selection of rules for business and organizations [38] under the General Data Protection Regulation (GDPR) [9] (shortened)

References

1. Dama International. "DAMA-DMBOK: Data Management Body of Knowledge (2nd Edition)". Technics Publications, LLC, Denville, NJ, USA, pp. 71, 2017.
2. Martens, Bertin and de Streel, Alexandre and Graef, Inge and Tombal, Thomas and Duch-Brown, Néstor, Business-to-Business Data Sharing: An Economic and Legal Analysis (July 22, 2020). EU Science Hub, 2020, Available at SSRN: https://ssrn.com/abstract=3658100
3. Pedreschi, D. et al. (2008). Privacy Protection: Regulations and Technologies, Opportunities and Threats. In: Giannotti, F., Pedreschi, D. (eds) Mobility, Data Mining and Privacy. Springer, Berlin, Heidelberg. https://doi.org/10.1007/978-3-540-75177-9_5
4. Seun Solomon Bakare, Adekunle Oyeyemi Adeniyi, Chidiogo Uzoamaka Akpuokwe, & Nkechi Emmanuella Eneh. (2024). Data Privacy Laws And Compliance: A Comparative Review Of The EU GDPR And USA Regulations. Computer Science & IT Research Journal, 5(3), 528–543. https://doi.org/10.51594/csitrj.v5i3.859
5. Custers, Bart and Bachlechner, Daniel. 'Advancing the EU Data Economy: Conditions for Realizing the Full Potential of Data Reuse'. 1 Jan. 2017: 291 – 309.

6. Regulation (EU) 2022/1925 of the European Parliament and of the Council of 14 September 2022 on contestable and fair markets in the digital sector and amending Directives (EU) 2019/1937 and (EU) 2020/1828 (Digital Markets Act) (Text with EEA relevance), available at https://eur-lex.europa.eu/legal-content/EN/TXT/?toc=OJ%3AL%3A2022%3A265%3ATOC&uri=uriserv%3AOJ.L_.2022.265.01.0001.01.ENG
7. Regulation (EU) 2022/2065 of the European Parliament and of the Council of 19 October 2022 on a Single Market For Digital Services and amending Directive 2000/31/EC (Digital Services Act) (Text with EEA relevance), available at https://eur-lex.europa.eu/legal-content/EN/TXT/?uri=CELEX%3A32022R2065
8. Regulation (EU) 2022/868 of the European Parliament and of the Council of 30 May 2022 on European data governance and amending Regulation (EU) 2018/1724 (Data Governance Act) (Text with EEA relevance), available at https://eur-lex.europa.eu/legal-content/EN/TXT/?uri=CELEX%3A32022R0868
9. Regulation (EU) 2016/679 of the European Parliament and of the Council of 27 April 2016 on the protection of natural persons with regard to the processing of personal data and on the free movement of such data, and repealing Directive 95/46/EC (General Data Protection Regulation) (Text with EEA relevance), https://eur-lex.europa.eu/eli/reg/2016/679/oj
10. Voigt, P., & Von dem Bussche, A. (2017). The eu general data protection regulation (gdpr). A Practical Guide, 1st Ed., Cham: Springer International Publishing, 10(3152676), 10–5555. https://doi.org/10.1007/978-3-319-57959-7
11. Ma, J. (2022). Emerging Digital Markets and Regulation. In: Regulating Data Monopolies. Springer, Singapore. https://doi.org/10.1007/978-981-16-8766-2_2
12. Heldt, A.P. (2022). EU Digital Services Act: The White Hope of Intermediary Regulation. In: Flew, T., Martin, F.R. (eds) Digital Platform Regulation. Palgrave Global Media Policy and Business. Palgrave Macmillan, Cham. https://doi.org/10.1007/978-3-030-95220-4_4
13. Datalagen (Data Act), SFS (Svensk Författningssamling; Swedish Code of Statutes) 1973:289
14. Riccardi JL (1983) The German Federal Data Protection Act of 1977: Protecting the right to privacy? Boston College International & Comparative Law Review 6(1): 243–272.
15. Directive 95/46/EC of the European Parliament and of the Council of 24 October 1995 on the protection of individuals with regard to the processing of personal data and on the free movement of such data http://data.europa.eu/eli/dir/1995/46/oj
16. Regulation (Eu) 2016/679 of the European Parliament and of the Council of 27 April 2016 on the protection of natural persons with regard to the processing of personal data and on the free movement of such data, and repealing Directive 95/46/EC (General Data Protection Regulation). http://data.europa.eu/eli/reg/2016/679/oj
17. Regulation (EU) 2022/2065 of the European Parliament and of the Council of 19 October 2022 on a Single Market For Digital Services and amending Directive 2000/31/EC (Digital Services Act) (Text with EEA relevance). http://data.europa.eu/eli/reg/2022/2065/oj
18. Regulation (EU) 2022/1925 of the European Parliament and of the Council of 14 September 2022 on contestable and fair markets in the digital sector and amending Directives (EU) 2019/1937 and (EU) 2020/1828 (Digital Markets Act) (Text with EEA relevance). http://data.europa.eu/eli/reg/2022/1925/oj
19. Regulation (EU) 2022/868 of the European Parliament and of the Council of 30 May 2022 on European data governance and amending Regulation (EU) 2018/1724 (Data Governance Act) (Text with EEA relevance). http://data.europa.eu/eli/reg/2022/868/oj
20. Regulation (EU) 2023/2854 of the European Parliament and of the Council of 13 December 2023 on harmonised rules on fair access to and use of data and amending Regulation (EU) 2017/2394 and Directive (EU) 2020/1828 (Data Act) (Text with EEA relevance). http://data.europa.eu/eli/reg/2023/2854/oj
21. Lonzetta AM, Hayajneh T. Challenges of Complying with Data Protection and Privacy Regulations. EAI Endorsed Scal Inf Syst [Internet]. 2020 Sep. 18 [cited 2024 Oct. 14];8(30):e4. Available from: https://publications.eai.eu/index.php/sis/article/view/2082
22. The Digital Services Act package, European Commission, available at https://digital-strategy.ec.europa.eu/en/policies/digital-services-act-package

23. The Digital Markets Act, European Commission, available at https://digital-markets-act.ec.europa.eu/index_en
24. About the Digital Markets Act, European Commission, available at https://digital-markets-act.ec.europa.eu/about-dma_en
25. DSA: Making the online world safer, European Commission, available at https://digital-strategy.ec.europa.eu/en/policies/safer-online
26. European Data Governance Act, European Commission, available at https://digital-strategy.ec.europa.eu/en/policies/data-governance-act
27. Directive (EU) 2019/1024 of the European Parliament and of the Council of 20 June 2019 on open data and the re-use of public sector information (recast), available at https://eur-lex.europa.eu/legal-content/EN/TXT/?qid=1561563110433&uri=CELEX:32019L1024
28. Data Governance Act explained, European Commission, available at https://digital-strategy.ec.europa.eu/en/policies/data-governance-act-explained
29. EU register of data intermediation services, European Commission, available at https://digital-strategy.ec.europa.eu/en/policies/data-intermediary-services
30. EU register of recognised data altruism organisations, European Commission, available at https://digital-strategy.ec.europa.eu/en/policies/data-altruism-organisations
31. European data strategy, European Commission, available at https://commission.europa.eu/strategy-and-policy/priorities-2019-2024/europe-fit-digital-age/european-data-strategy_en#:~:text=The%20European%20data%20strategy%20aims,businesses%2C%20researchers%20and%20public%20administrations
32. Regulation (EU) 2023/2854 of the European Parliament and of the Council of 13 December 2023 on harmonised rules on fair access to and use of data and amending Regulation (EU) 2017/2394 and Directive (EU) 2020/1828 (Data Act) (Text with EEA relevance), available at https://eur-lex.europa.eu/eli/reg/2023/2854
33. Data Act explained, European Commission, available at https://digital-strategy.ec.europa.eu/en/factpages/data-act-explained
34. HLF Workstream 14 on Data Interoperability - Final Report, available at https://ec.europa.eu/docsroom/documents/58914
35. Directive 95/46/EC of the European Parliament and of the Council of 24 October 1995 on the protection of individuals with regard to the processing of personal data and on the free movement of such data, available at https://eur-lex.europa.eu/legal-content/EN/TXT/?uri=celex%3A31995L0046
36. The California Consumer Privacy Act (CCPA), available at https://oag.ca.gov/privacy/ccpa
37. What does the General Data Protection Regulation (GDPR) govern?, European Commission, available at https://commission.europa.eu/law/law-topic/data-protection/reform/what-does-general-data-protection-regulation-gdpr-govern_en
38. Reform of EU data protection rules, European Commission, available at https://commission.europa.eu/law/law-topic/data-protection/reform_en
39. Alessandro Mantelero, Giuseppe Vaciago, Maria Samantha Esposito, Nicole Monte, The common EU approach to personal data and cybersecurity regulation, International Journal of Law and Information Technology, Volume 28, Issue 4, Winter 2020, Pages 297–328, https://doi.org/10.1093/ijlit/eaaa021

Chapter 9
Introduction to the European Machinery Regulation

Christophe Gaie, Markus Mueck, and Christophe Gossard

Abstract In June 2023, the European Union adopted Regulation 2023/1230/EU on Machinery (Machinery Regulation). This regulation replaces Directive 2006/42/EC and sets forth essential health and safety requirements for machinery placed on the European Union (EU) Single Market. Similar to the far-reaching impact expected from the EU AI Act, the Machinery Regulation is anticipated to significantly influence machinery safety standards globally. As such, the new revision of Directive 2006/42/EC addresses novel technologies which challenge product safety in new ways. Indeed, artificial intelligence, the Internet of things and robotics are rather new to the world of machinery and inherent safety protocols need to be reconsidered correspondingly. For example, based on the new tools machinery products are able to let decision making functionalities rely on real-time processing of information, problem solving, mobility, sensor systems, learning, adaptability, and capability of operating in unstructured environments. This chapter provides an overview of the Machinery Regulation, including the institutions responsible for its implementation and enforcement. It explores how manufacturers can achieve compliance with the regulation's requirements. The chapter categorizes machinery based on risk, explaining how these categories influence the specific safety measures needed. Finally, it examines exemptions for specific machinery types and the importance of harmonized standards in demonstrating compliance.

Keywords Regulation · Machinery safety · European Union single market · Risk assessment · Harmonized standards

---

C. Gaie (✉)
Head of the Engineering and Digital Innovation Division, Prime Minister Services, Paris, France
e-mail: Christophe.Gaie@gmail.com

M. Mueck
Intel Deutschland GmbH, Unterhaching, Germany
e-mail: Markus.Dominik.Mueck@intel.com

C. Gossard
John Deere, Mannheim, Germany
e-mail: Gossardchristophe@johndeere.com

M. Mueck and C. Gaie (eds.), *European Digital Regulations*, Intelligent Systems Reference Library 265, https://doi.org/10.1007/978-3-031-80809-8_9

9.1 Introduction

9.1.1 Historical Overview

The Machinery Regulation builds upon a long history of EU efforts to protect workers and consumers from machinery-related hazards. Since the early twentieth century, the EU has implemented various directives and regulations to address evolving safety concerns [1, 2]. However, the rapid advancements in technology, particularly in areas like artificial intelligence and the Internet of Things, have necessitated a more comprehensive and flexible framework. This evolution is depicted in Fig. 9.1 and explained below.

First, the Council Directive 87/357/EEC of 25 June 1987 aimed to protect consumers and ensure the free movement of safe goods within the EU by harmonizing laws across member states regarding products that mislead consumers by appearing to be food but are not. This was particularly important for protecting children, who may be especially vulnerable to these misleading products. By eliminating trade barriers caused by differing national laws, the directive helped to create a level playing field for businesses and promote competition while also safeguarding consumer safety.

Then, the Council Directive 92/59/EEC established a comprehensive framework for general product safety within the EU. This directive sought to enhance consumer protection by ensuring member states had robust systems in place to identify and mitigate risks associated with consumer products. Furthermore, it aimed to facilitate the market access of safe products by promoting harmonized safety standards across the EU.

Thereafter, the Directive 2001/95/EC, which superseded Directive 92/59/EEC, aimed to further strengthen the general product safety framework. Key points of Directive 2001/95/EC included strengthened producer responsibility for product safety, enhanced market surveillance and information exchange about unsafe products, and improved procedures for product recalls.

Next, the Directive (EU) 2020/1828 of the European Parliament and of the Council of 25 November 2020 focuses on enhancing consumer protection through collective

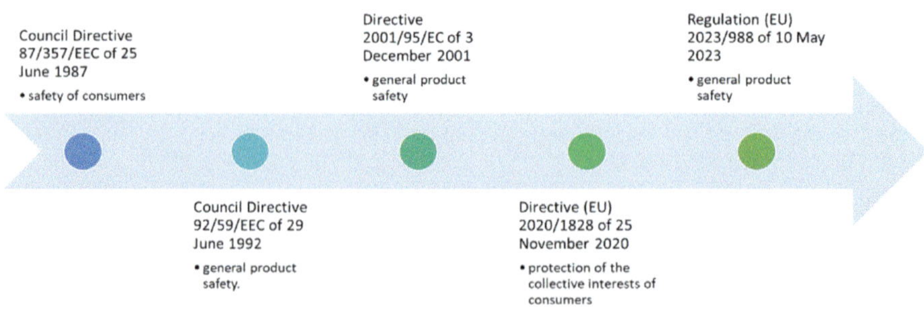

Fig. 9.1 Timeline of the EU action to establish product safety

redress mechanisms. It empowers consumer organizations to represent a group of consumers who have been negatively impacted by unfair business practices, enabling them to pursue legal action collectively. This directive aims to address imbalances in consumer-business relationships and provide consumers with a more effective means of seeking justice.

Finally, the most recent regulation, (EU) 2023/988, modernizes product safety requirements to reflect technological advancements. It strengthens online market surveillance and streamlines product safety assessment procedures, further enhancing the overall product safety framework in the EU. This regulation supersedes both Directives 87/357/EEC and 2001/95/EC, consolidating the legal framework for general product safety.

The authors outline the fact that the Machinery Regulation (EU) 2023/988 repeals the previous directives, making it the current legal framework for general product safety in the EU. It consolidates existing requirements and addresses emerging challenges, it aims to create a safer and more innovative machinery landscape. This is particularly important given the increasing globalization of the machinery industry and the need for harmonized standards across different regions (Europe, Asia, North America, etc.).

9.1.2 Context of the Machinery Directive

For decades, the European Union (EU) has prioritized ensuring the safety of individuals interacting with machinery within its borders. This commitment is reflected in the recently adopted Regulation 2023/1230/EU on Machinery (Machinery Regulation) [3]. This regulation, effective in June 2023, replaces the longstanding Machinery Directive (2006/42/EC) [4]. The present chapter delves into the Machinery Regulation, offering a comprehensive guide for navigating the new landscape.

Prior to the Machinery Regulation, the Machinery Directive (2006/42/EC) served as the cornerstone of machinery safety regulations in the EU. Essential health and safety requirements are established by this Directive for machinery placed on the market. However, the evolving technological landscape necessitated a more comprehensive framework.

First, the Machinery Regulation clarifies two major and complex issues: "*incomplete machinery*"—also addressed as "*partly-completed machinery*" [5]—and "*substantial modifications*" [6].

Second, the Machinery Regulation addresses advancements in new technologies especially autonomous mobile machinery (robots), Internet of Things (IoT) and Artificial intelligence (AI). While retaining the core safety principles of the directive, the Machinery Regulation offers a more robust framework for ensuring machinery safety in the face of these new technological realities. These novelties are illustrated in Fig. 9.2.

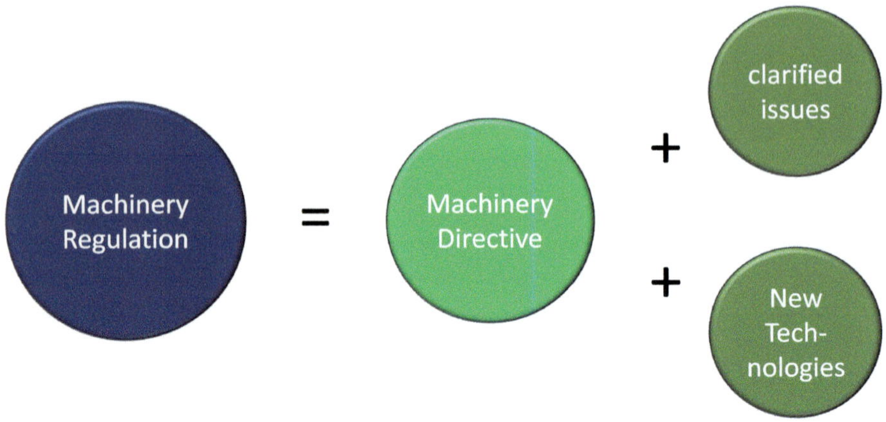

Fig. 9.2 Overview of the machinery regulation

The Machinery Regulation aims to ensure that machinery, related products and partly completed machinery are delivered on the European market while protecting the health and safety of persons, consumers as well as professional users.

9.2 Scope and Exemptions

The Machinery Regulation encompasses a wide range of machinery, promoting safety across various industries with the following classification [3, 7]:

- **Machinery**: it includes any device with moving parts or connected to other machinery, used for processing, handling, moving, or treating materials. This corresponds, for example, to industrial robots, machine tools, and packaging machinery.
- **Related Products**: this concerns interchangeable equipment, safety components (like pressure relief valves), lifting accessories (slings and hooks), chains, ropes, and removable mechanical transmission devices. These products directly contribute to the overall safety of machinery.
- **Partly Completed Machinery**: this refers to machinery intentionally designed for incorporation into a larger machinery system before final placement on the market.

The formalization of these categories ensures a comprehensive and understandable approach to safety, considering not just the final assembled machinery but also its individual components and potential configurations. While the Machinery Regulation aims for broad coverage, specific equipment categories are excluded [3, 8]:

- **Motor Vehicles and Trailers**: These are subject to separate regulations focused on roadworthiness and safety, see Regulation (EU) 2019/2144 [9].

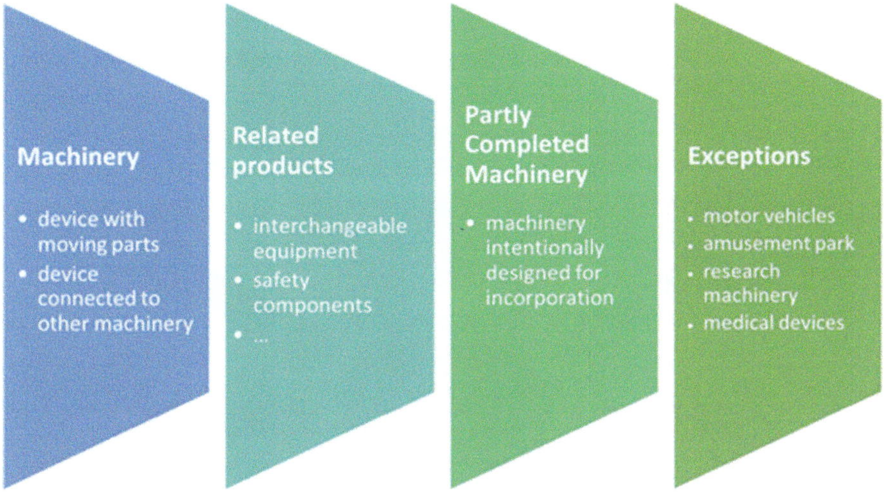

Fig. 9.3 Overview of the scope and exemptions of the regulation

- **Fairground and Amusement Park Equipment**: Machinery designed solely for entertainment purposes in these settings falls outside the scope.
- **Research Machinery for Temporary Use in Laboratories**: Prototypes and research-specific machinery used for a limited time in controlled laboratory environments are exempt.
- **Medical Devices and Lifts**: These categories have dedicated regulations addressing their specific safety requirements, see Regulation (EU) 2017/745 [10] and Directive 2014/33/EU [11].

Manufacturers will have to carefully verify whether their equipment fulfills the exemptions listed in the Machinery Regulation. In case of doubt, they are advised to contact the relevant authorities to check the eligibility of their equipment. By clearly defining the scope and exemptions, the Machinery Regulation ensures its focus remains on machinery with the potential to impact safety in the EU market. This is illustrated in Fig. 9.3.

9.3 Essential Health and Safety Requirements

A key focus of the Machinery Regulation is the introduction of essential requirements for health and safety, outlined in Annex III. These requirements define the requirements that manufacturers must follow to design and construct machinery that prioritizes user safety [3, 12] and are illustrated in Fig. 9.4.

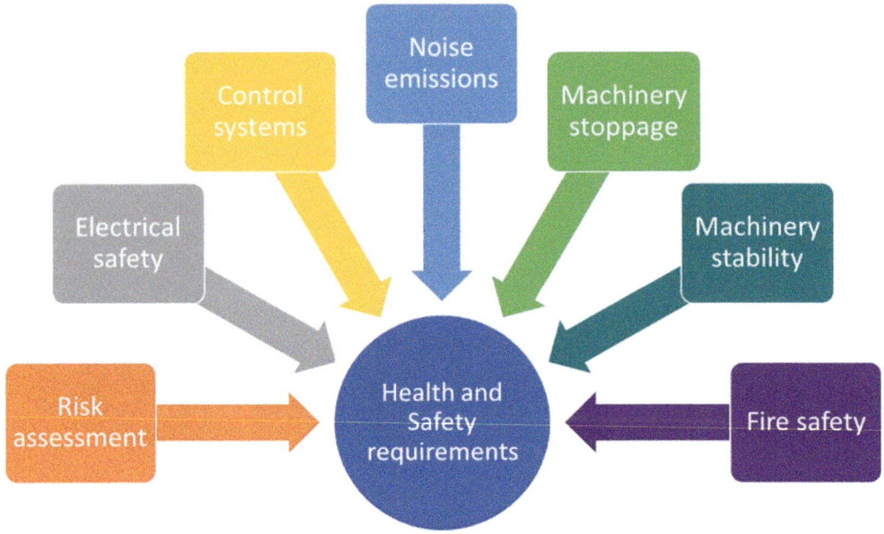

Fig. 9.4 Main consideration to ensure health and safety

Manufacturers are obliged to meet Essential Health and Safety Requirements to achieve access to the European Union Single Market and if available, harmonized standards are associated with these Essential Health and Safety Requirements (EHSRs). Typically, an annex correlates the specific clauses in EN harmonized standards with the EHSRs of the Regulation.

First, the regulation emphasizes inherent safety measures integrated into the machinery design. This includes minimizing risks through features like safeguarding, stability, and ergonomic design. Then, manufacturers are required to conduct a thorough risk assessment to identify potential hazards associated with their machinery and implement appropriate control measures to mitigate those risks.

Furthermore, the regulation sets strict requirements for electrical installations within machinery, minimizing the risk of electrical shock and fire. Clear and user-friendly control systems are also mandated, ensuring operators can safely operate and interact with the machinery. Moreover, machinery noise levels must be kept within acceptable limits to protect users from hearing damage. Mechanisms for safe and controlled stopping of machinery are also required, preventing accidental operation or injuries during maintenance.

Finally, the regulation emphasizes stable machinery design to prevent tipping or overturning, ensuring operator safety. There are also some measures to minimize the risk of fire hazards within machinery are outlined, promoting a safe working environment. It's important to underline that Annex III provides a general framework. Specific requirements may vary depending on the type and complexity of the machinery.

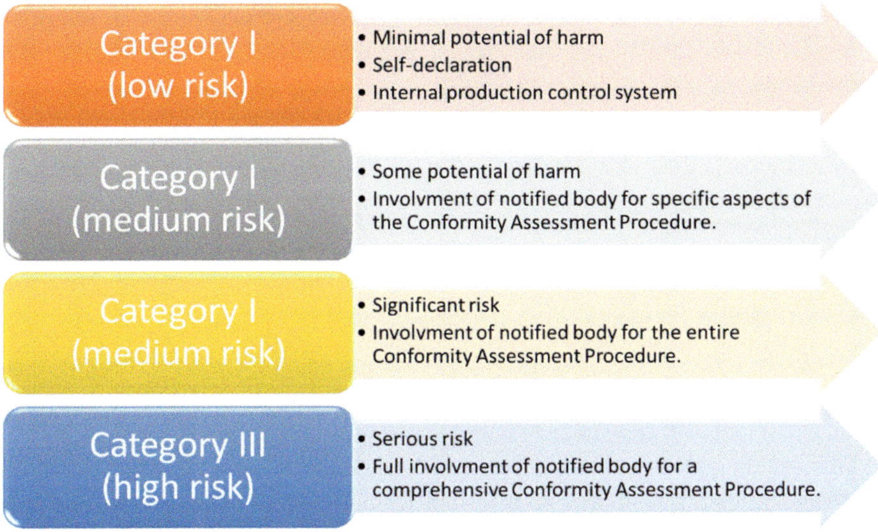

Fig. 9.5 Correspondence between categories and conformity assessment procedures

9.4 Conformity Assessment Procedures

9.4.1 Risk Categories and Conformity Assessment Procedures (Annex I)

The Machinery Regulation establishes a tiered system for demonstrating adherence to the essential safety requirements outlined in Annex III. This system is based on risk categories assigned to different types of machinery (defined in Annex I) and is illustrated in Fig. 9.5. The higher the risk posed by the machinery, the more rigorous the Conformity Assessment Procedure (CAP) required [3, 13].

First, Category I (Low Risk) concerns machinery with minimal risk, typically requiring the manufacturer to declare conformity through an internal production control system (explained in Sect. 4.2).

Second, Category II 1 (Medium Risk) refers to machinery with some potential for harm, often requiring involvement of a notified body[1] for specific aspects of the Conformity Assessment Procedure.

Then, Category II 2 (Medium Risk) addresses machinery with more significant risk, typically involving a notified body for the entire Conformity Assessment Procedure.

[1] A notified body is an independent organization designated by an EU member state to assess the conformity of certain products before they are placed on the EU market.

Finally, Category III (High Risk) refers to machinery posing serious risk, requiring a comprehensive Conformity Assessment Procedure with full involvement of a notified body.

It is important to underline that identifying the risk category of the machinery is crucial for selecting the appropriate Conformity Assessment Procedure.

9.4.2 Internal Production Control (Annex VI)

Internal Production Control (IPC) is a Conformity Assessment Procedure applicable to Category I machinery (low risk). Under IPC, the manufacturer takes full responsibility for demonstrating adherence to the essential safety requirements through a documented and well-maintained quality management system.

This system typically involves regular risk assessments for the machinery, design and manufacturing processes that ensure consistent safety standards, technical documentation outlining the design, production, and operation of the machinery as well as internal audits and inspections to verify compliance with established procedures.

By implementing a robust IPC system, manufacturers can ensure their low-risk machinery meets the safety requirements of the Machinery Regulation.

9.4.3 Third-Party Conformity Assessment Procedures (Annexes VII-X)

For **machinery in Categories II 1, II 2, and III** (medium to high risk), the Machinery Regulation mandates involvement of a **notified body**. These are independent organizations designated by EU member states to assess conformity with the regulation. Depending on the risk category, different third-party Conformity Assessment Procedures are available (Annexes VII–X):

- **Annex VII (EU-Type Examination)**: A notified body examines the technical documentation and a prototype of the machinery to verify conformity. This is typically used for Category II 1 machinery.
- **Annex VIII (Conformity to Type based on Internal Production Control)**: Combines internal production control with notified body involvement for specific aspects like final product inspections. This applies to some Category II 1 and II 2 machinery.
- **Annex IX (Conformity Based on Full Quality Assurance)**: A comprehensive procedure where a notified body closely monitors the manufacturer's entire production process to ensure consistent conformity. This is typically used for Category II 2 and III machinery.

- **Annex X (Conformity Based on Unit Verification)**: A notified body assesses each individual unit of high-risk (Category III) machinery before it can be placed on the market.

By utilizing the appropriate third-party Conformity Assessment Procedure, manufacturers of medium to high-risk machinery can demonstrate adherence to the regulation and obtain the necessary approvals for market access within the EU.

9.5 CE Marking and User Instructions

The Machinery Regulation aims simultaneously to offer security to European citizens and establish a dynamic economic environment. To this end, the CE Marking offers an easy and effective build up confidence between companies, consumers and government authorities.

Once manufacturers have successfully demonstrated conformity of their machinery to the essential safety requirements through the appropriate Conformity Assessment Procedure, they can affix the CE marking and provide clear user instructions. These two elements are crucial for placing machinery on the EU market.

9.5.1 Obtaining the CE Marking

The **CE marking** is a mandatory declaration by the manufacturer that their machinery complies with all applicable EU requirements, including the Machinery Regulation. It is a visual symbol that allows free movement of products within the EU single market. To affix the CE marking, manufacturers must follow the process of Fig. 9.6:

- **Complete the Conformity Assessment Procedure**: Successfully undergo the required conformity assessment process according to the machinery's risk category. *Reference: Article 16 of the Machinery Regulation.*
- **Draw Up the EU Declaration of Conformity**: This document formally declares compliance with the Machinery Regulation and includes details about the conformity assessment procedure used. *Reference: Article 17 of the Machinery Regulation.*
- **Create Technical Documentation**: Prepare a comprehensive technical file documenting the design, manufacturing, and testing processes of the machinery. *Reference: Article 12 of the Machinery Regulation.*
- **Verify Machinery Compliance**: Ensure that the final product meets the specifications outlined in the technical documentation and the essential safety requirements.

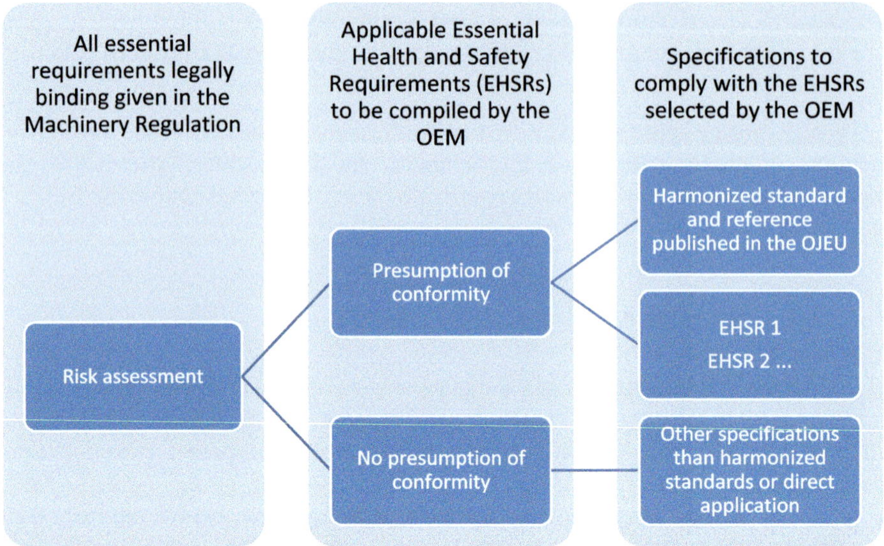

Fig. 9.6 Summary of the conformity assessment procedure considering applicable Essential Health and Safety Requirements (EHSRs)

The CE marking should be affixed to the machinery in a visible, legible, and indelible manner, allowing for easy identification. The full process is illustrated in Fig. 9.7.

9.5.2 User Instructions and Digital Formats

The Machinery Regulation outlines that clear and comprehensive user instructions are essential for safe and efficient machinery operation. It also mandates that manufacturers provide user instructions along with the machinery.

These instructions should be written in a clear and understandable language, accessible to the intended user. For example, the manual of a machine sold in France should be written in French. Key information to include in the user instructions are details below:

- **Safe installation and assembly**: Instructions on how to safely set up and assemble the machinery;
- **Safe operation**: Detailed guidance on how to operate the machinery correctly, including safety precautions and emergency procedures;
- **Maintenance and cleaning**: Instructions on regular maintenance and cleaning to ensure continued safe operation;
- **Troubleshooting**: Information on common issues and how to resolve them.

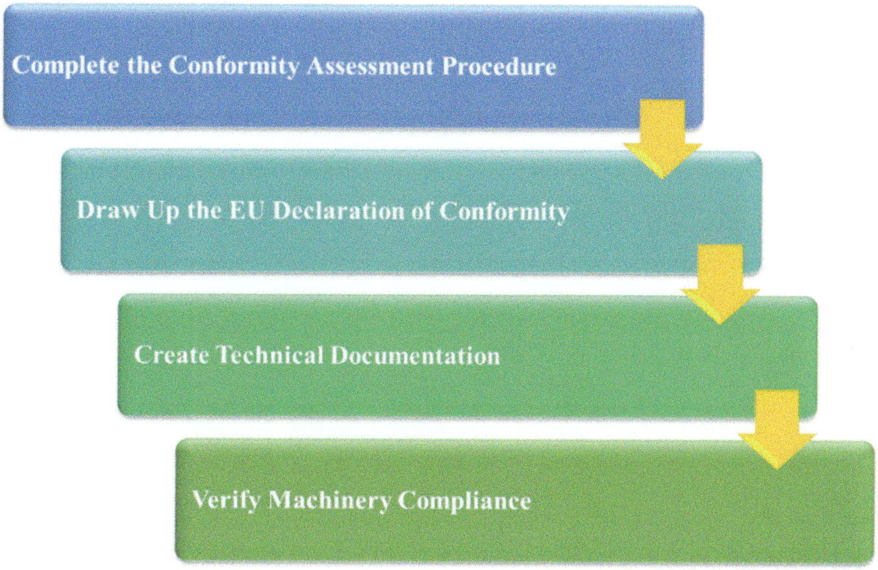

Fig. 9.7 Steps to obtain the CE marking

The Machinery Regulation allows for **digital formats** for user instructions, provided they are easily accessible to the user. However, manufacturers must ensure that users have the necessary hardware and software to access and read the digital instructions. *Reference: Article 14 of the Machinery Regulation.*

By providing clear and accessible user instructions, manufacturers contribute to safe machinery operation and minimize the risk of accidents.

9.6 Roles and Responsibilities

Successful implementation of the Machinery Regulation hinges on clear roles and responsibilities for various stakeholders involved in the machinery lifecycle. This sections details Manufacturers' Obligations as well as Notified Bodies and National Authorities.

9.6.1 Manufacturers' Obligations

The responsibility for ensuring that machinery complies with the Regulation is assumed by manufacturers and imply the following activities, also illustrated in Fig. 9.8:

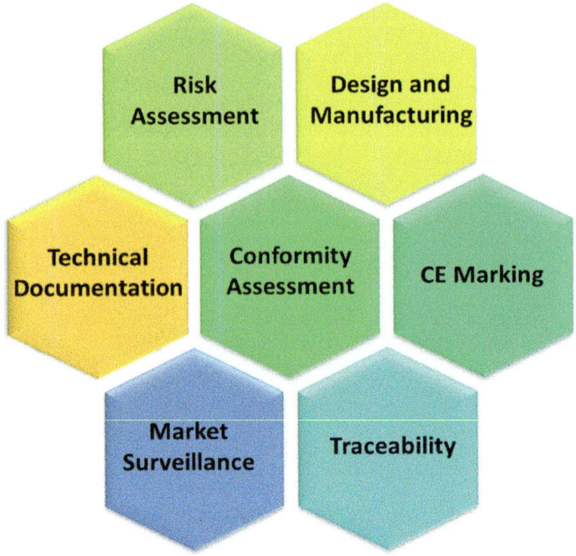

Fig. 9.8 Summary of manufacturers' obligations

- **Risk Assessment**: Manufacturers have to achieve a thorough risk assessment to identify potential hazards and implement appropriate safety measures. *Reference: Article 5 of the Machinery Regulation.*
- **Design and Manufacturing**: Manufacturers also are in charge of ensuring that machinery is designed and manufactured in accordance with the essential safety requirements. *Reference: Article 6 of the Machinery Regulation.*
- **Technical Documentation**: Manufacturers must provide a technical documentation detailing the design, manufacturing, and testing processes of the machinery. *Reference: Article 12 of the Machinery Regulation.*
- **Conformity Assessment**: Manufacturers are in charge of selecting and implementing the appropriate Conformity Assessment Procedure based on the machinery's risk category. *Reference: Article 16 of the Machinery Regulation.*
- **CE Marking**: Manufacturers are expected to attach the CE marking to the machinery after successful completion of the Conformity Assessment Procedure. *Reference: Article 17 of the Machinery Regulation.*
- **Market Surveillance**: Manufacturers are expected to collaborate with market surveillance authorities to address any issues or non-conformities identified after the machinery is placed on the market. *Reference: Article 21 of the Machinery Regulation.*
- **Traceability**: Finally, manufacturers are assigned to maintaining records that allow for tracing the machinery throughout its lifecycle. *Reference: Article 18 of the Machinery Regulation.*

Thus, manufacturers should establish robust quality management systems to ensure ongoing compliance with the Machinery Regulation and address any potential issues proactively.

9.6.2 Notified Bodies and National Authorities

The role of notified bodies and national authorities is central to ensure the effectivity of the Machinery Regulation. Both of them are described in the two following paragraphs and are illustrated in Fig. 9.9:

Notified Bodies:

- **Conformity Assessment**: Notified bodies are in charge of conducting conformity assessments for machinery in categories II 1, II 2, and III, as required by the Machinery Regulation. *Reference: Article 16 of the Machinery Regulation.*
- **Issuing Certificates**: Notified bodies have to provide certificates to manufacturers whose machinery successfully passes the conformity assessment. *Reference: Article 25 of the Machinery Regulation.*
- **Monitoring Manufacturers**: Notified bodies are in charge of conducting surveillance activities to ensure continued compliance. *Reference: Article 27 of the Machinery Regulation.*
- **Technical Expertise**: Notified bodies have to provide technical expertise and support to manufacturers and market surveillance authorities.

National Authorities:

- **Designation of Notified Bodies**: National authorities are in charge of designating competent organizations as notified bodies. *Reference: Article 24 of the Machinery Regulation.*
- **Market Surveillance**: National authorities have to monitor the market to ensure compliance with the Machinery Regulation and taking corrective actions when necessary. *Reference: Article 21 of the Machinery Regulation.*
- **Enforcement**: National authorities have to enforce the Machinery Regulation and imposing penalties for non-compliance. *Reference: Article 30 of the Machinery Regulation.*

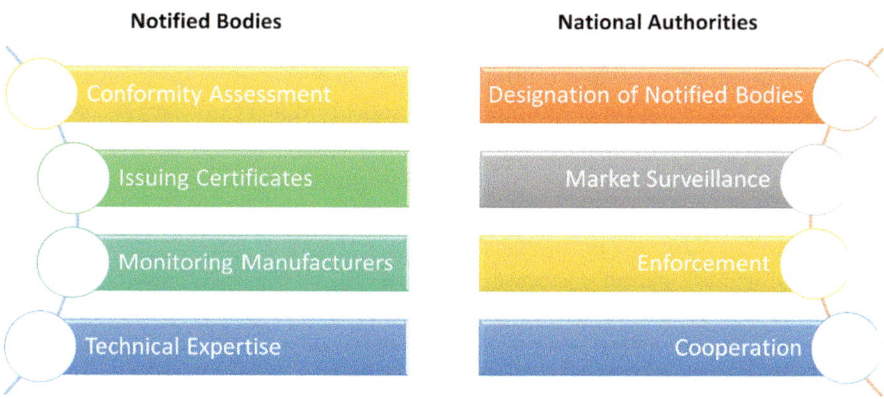

Fig. 9.9 Role of notified bodies and national authorities

- **Cooperation**: National authorities must collaborate with other EU member states on market surveillance and enforcement activities.

It is important to notice that effective cooperation between manufacturers, notified bodies, and national authorities is essential for ensuring the safety of machinery within the EU market.

9.7 Implementation Timeline and Application Considerations

9.7.1 Implementation Timeline

The Machinery Regulation follows a phased implementation timeline to allow manufacturers and other stakeholders to adapt to the new requirements. There are three important dates to outline, as illustrated in Fig. 9.10:

- **June 14, 2023**: The Machinery Regulation was published in the Official Journal of the European Union, marking the official start of the implementation process. *Reference: Official Journal of the European Union, L 168/1.*
- **January 14, 2024**: Certain provisions related to notified bodies became effective, allowing for their designation and accreditation. *Reference: Article 24 of the Machinery Regulation.*
- **January 20, 2027**: The full application of the Machinery Regulation commences, with all requirements becoming mandatory for machinery placed on the market. *Reference: Article 45 of the Machinery Regulation.*

Three phases of implementations are defined and described below:

1. **Preparation and Adaptation**

   During this first phase, manufacturers will familiarize themselves with the Machinery Regulation and assess the impact on their existing machinery. In the meantime, notified bodies will undergo the necessary accreditation processes to be able to conduct conformity assessments. *Reference: Article 24 of the Machinery Regulation.* It is also required that member states designate notified

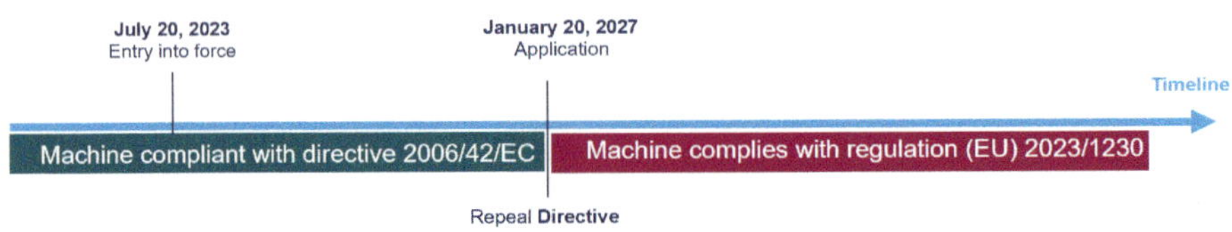

Fig. 9.10 Machinery directive implementation timeline

bodies and establish market surveillance systems. *Reference: Article 24 of the Machinery Regulation.*

2. **Transition Period**

   An intermediate period is scheduled between the publication of the regulation and its full application to enable a gradual transition. During this phase, manufacturers can continue to place machinery on the market according to the provisions of the Machinery Directive (2006/42/EC) until January 20, 2027. However, it is advisable for manufacturers to start preparing for the new requirements during this period.

3. **Full Application**

   Finally, the Regulation will be fully applied on January 20, 2027. From this date, each machinery placed on the EU market will comply with the Machinery Regulation.

This means that manufacturers must implement the necessary changes to their production processes and quality management systems. Moreover, notified bodies have to be fully operational in conducting conformity assessments and market surveillance authorities will enforce the regulation rigorously.

9.7.2 Application Considerations

To facilitate the application of the Machinery Regulation, the following adaptations are defined:

- **Existing Machinery**: Machinery already on the market before January 20, 2027, is not subject to the new requirements. However, any substantial modifications made after this date will need to comply with the Machinery Regulation. *Reference: Article 3 of the Machinery Regulation.*
- **Harmonized Standards**: Manufacturers are encouraged to use harmonized standards to facilitate compliance. It is essential to keep up-to-date with the development of new standards related to the Machinery Regulation. *Reference: Article 5 of the Machinery Regulation.*
- **Continuous Improvement**: The Machinery Regulation is a dynamic framework. Manufacturers should continuously monitor developments and updates to ensure ongoing compliance.

9.8 Standards in Support of the Machinery Regulation

The Machinery Regulation is governed under the New Legislative Framework (NLF) and thus relies on Harmonised European Norms (or Harmonised Standards) for its implementation in certain cases (see section X.4 on Conformity Assessment above).

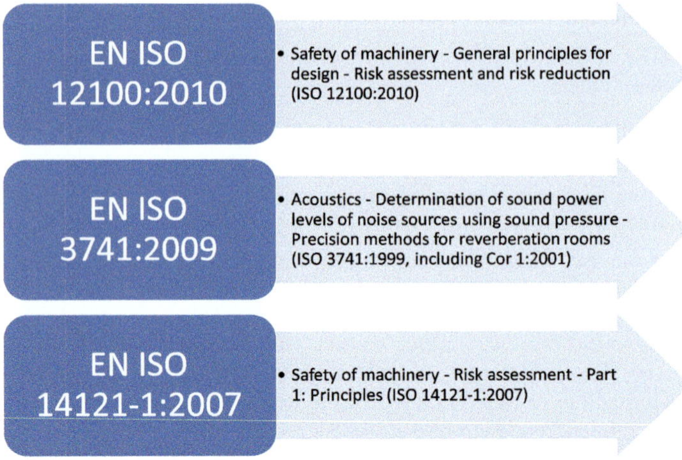

Fig. 9.11 Selection of Harmonised Standards in support of machinery regulation and directive

European Standardisation Organisations (ESOs) CEN/CENELEC have indeed been tasked to develop such Harmonised Standards in support the previously existing Machinery Directive 2006/42/EC already. The European Commission publishes corresponding list of available Harmonised Standards on its website [13]. The list is quite extensive, with some selected key standards summarized in Fig. 9.11.

At the time when the present chapter was written, the European Commission was in the process of preparing a Standardisation Request (SR), currently available as a Draft [14], triggering a review of the existing Harmonised Standards together with other Standards deliverables and the creation of new deliverables, especially in the space of new technologies including AI, Software, etc. They are illustrated in Fig. 9.12 and its timeline in Fig. 9.13.

Stakeholders are able to influence the development of the Standards Deliverables through participation in the CEN/CENELEC activities. It is highly recommended that all involved stakeholders use the opportunity for participation. This is the only way to ensure that the finished Standards Deliverables, especially the Harmonised European Norms—or Harmonised Standards—are written in accordance with the needs of relevant stakeholders including manufacturers.

9.9 Safety Components as a Key Concept of the Machinery Regulation

The concept of safety components has already been introduced in the prior Machinery Directive 2006/42/EC [4], but with a very limited scope as illustrated in Fig. 9.14.

Fig. 9.12 New standards deliverables as requested in the currently available draft standardistion request [14]

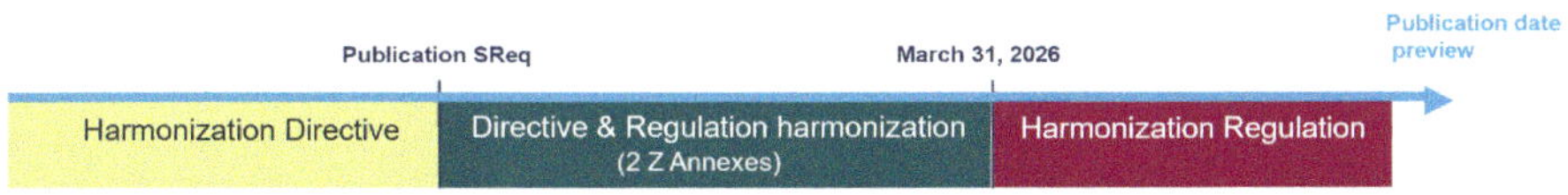

Fig. 9.13 Machinery directive implementation timeline including standardisation request

While traditional machinery equipment focused primarily on physical components, the Machinery Regulation [3] recognizes the growing role of digital technology and software in the industry. It suggests that the definition of safety components should encompass not only physical devices but also digital ones. This includes software that performs a safety function and is sold independently, which should be considered a safety component to reflect the increasing use of software in machinery design.

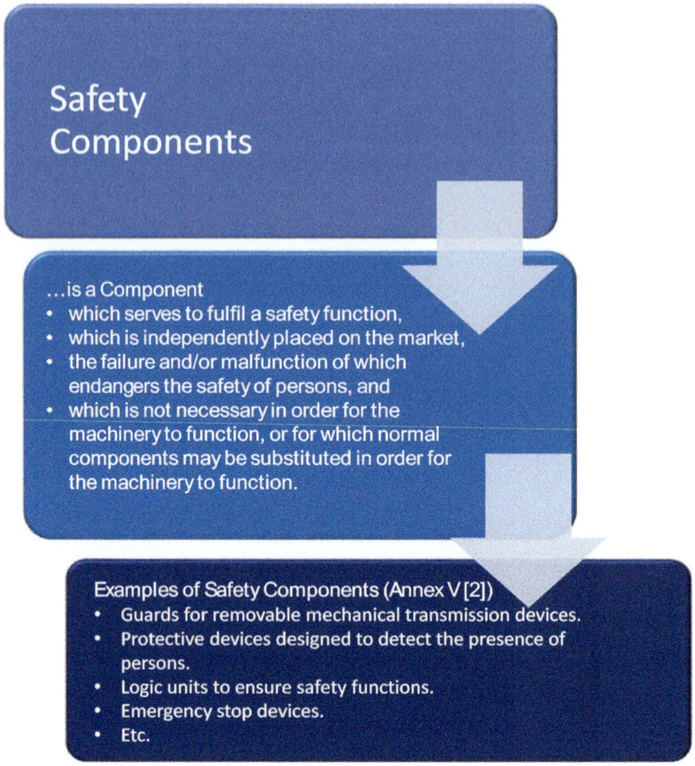

Fig. 9.14 Definition of a safety components by the prior Machinery Directive 2006/42/EC [4]

Consequently, a definition of a Safety Component is introduced in Table 9.1 that covers classical aspects as well as new technologies:

The inclusion of new technologies in the upper definition leads to an extended Indicative List of Safety Components (Annex II of the Machinery Regulation [3]). Specifically, notable extensions are indicated in Table 9.2.

This Annex is very important for the integration of software in safety components. The concept of self-evolving behavior necessitates third-party certification according to the Machinery Regulation. In contrast, for software that is released and remains unchanged, self-certification is permissible since it is not considered to be a Safety Component according to Annex II of the Machinery Regulation. For High Risk AI

Table 9.1 Definition of a safety component by the machinery regulation [3]

'Safety component' means a physical or digital component, including software, of a product within the scope of this Regulation, which is designed or intended to fulfil a safety function and which is independently placed on the market, the failure or malfunction of which endanger the safety of persons, but which is not necessary in order for that product to function or for which normal components may be substituted in order for that product to function

Table 9.2 Indicative list of safety components (Annex II of the machinery regulation [3])

...
18. Software ensuring safety functions
19. Safety components with fully or partially self-evolving behaviour using machine learning approaches ensuring safety functions
...

systems relating to specific product legislations, the EU AI Act [15] delegates the responsibility of determining which systems require this certification to the most relevant product legislation, and Annex I of the EU AI Act [15] lists specific product applicable legislations. This includes the Machinery Regulation (EU) 2023/1230.

The Machinery Regulation [3] requires rigorous third-party conformity assessments for safety components, including those with self-evolving safety functions. This applies regardless of whether the component is sold separately or integrated into machinery.

9.10 Interdependencies with Other European Regulations

In terms of Cybersecurity, the Machinery Regulation [3] specifically refers to the European Cybersecurity Act (Regulation (EU) 2019/881) [16].

As stated in the Machinery Regulation, to mitigate risks posed by malicious third-party attacks on product safety, this Regulation should incorporate essential health and safety requirements. Compliance with these requirements can be assumed if a product holds a cybersecurity certificate issued under a scheme adopted according to Regulation (EU) 2019/881.

Indeed, as further detailed in the respective chapter on the Cyber Security Act of the present book, Cybersecurity Certification Schemes are being developed under the Cybersecurity Act [16] under control of the European Union Agency for Cybersecurity (ENISA). Existing examples of such Cybersecurity Certification Schemes are proposed in Fig. 9.15 as follows.

The Machinery Regulation [3] further explains how presumption of conformity can be achieved using an applicable Cybersecurity Certification Scheme under the European Cybersecurity Act (Regulation (EU) 2019/881) [16].

Machinery and related products certified or declared compliant under a cybersecurity certification scheme adopted according to Regulation (EU) 2019/881 (and listed in the Official Journal) are presumed to meet the essential health and safety requirements of Annex III, sections 1.1.9 and 1.2.1, regarding protection against corruption and the safety and reliability of control systems, to the extent covered by the cybersecurity certificate or declaration.

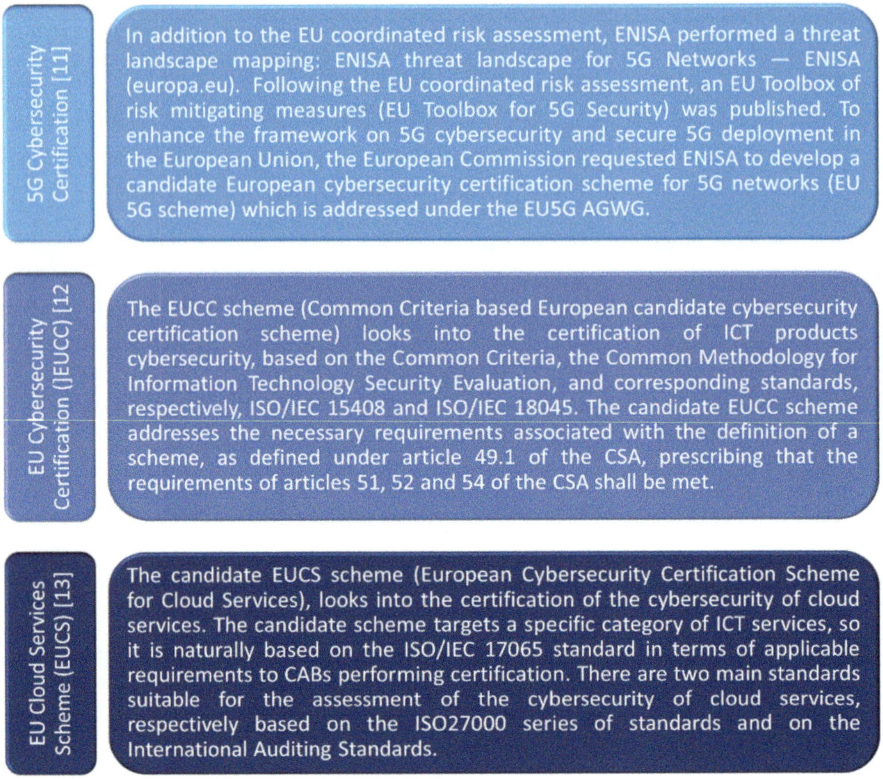

Fig. 9.15 Cybersecurity certification schemes under the European cybersecurity act (Regulation (EU) 2019/881) [16]

9.11 Conclusion

The Machinery Regulation 2023/1230/EU is a substantial new development going well beyond the previous Machinery Directive 2006/42/EC which was applicable since 2006. Especially, new technological development related to Artificial Intelligence, Cybersecurity and others have been included and a bridge to related other Regulations is made, notable to the Cyber Security Act [16]. Machinery manufacturers will thus need to develop their competences in those new fields.

Disclaimer: Note that the authors have assembled the information above to the best of their knowledge as available at the time when the present book was written. Any interpretation is the personal opinion of the authors. Also, the regulation framework can change and evolve any time. The only legally binding text is the one provided in the documents formally published in the European Union Official Journal.

References

1. Directive 2014/33/EU of the European Parliament and of the Council of 26 February 2014 on the harmonisation of the laws of the Member States relating to lifts and safety components for lifts (recast) Text with EEA relevance, available at https://eur-lex.europa.eu/eli/dir/2014/33/oj.
2. Ruohonen, J. A review of product safety regulations in the European Union. Int. Cybersecur. Law Rev. 3, 345–366 (2022). https://doi.org/10.1365/s43439-022-00057-8.
3. "Regulation (EU) 2023/1230 of the European Parliament and of the Council of 14 June 2023 on machinery and repealing Directive 2006/42/EC of the European Parliament and of the Council and Council Directive 73/361/EEC", Last visited on 15th July 2024, available at https://eur-lex.europa.eu/eli/reg/2023/1230/oj.
4. "DIRECTIVE 2006/42/EC OF THE EUROPEAN PARLIAMENT AND OF THE COUNCIL of 17 May 2006 on machinery, and amending Directive 95/16/EC,", Last visited on 15th July 2024, available at https://eur-lex.europa.eu/legal-content/EN/TXT/?uri=CELEX:02006L0042-20190726.
5. Advolo. "Did I make a Machine or a Partly-Completed Machinery? - AdVolo." AdVolo, November 28, 2023. https://www.advolo.it/en/news/did-i-make-a-machine-or-a-partly-completed-machinery/, Last visited on 15th July 2024.
6. Gt-Engineering. "SUBSTANTIAL MODIFICATIONS," October 1, 2023. https://www.gt-engineering.it/en/insights/machinery-directive/machinery-substantial-modifications/, Last visited on 15th July 2024.
7. Darren Hugheston-Roberts, 20 February 2024, "CDA Magazine - Machinery regulation gets an update," https://www.controlsdrivesautomation.com/Machinery-regulation-gets-an-update, Last visited on 15th July 2024.
8. Pierre Belingard, Stefano Boy, "ETUI Machinery: From the Directive to the new Regulation, what changes? - Eurogip-182/E - 07/2023" https://eurogip.fr/wp-content/uploads/2023/07/EUROGIP-ETUI_Machinery-from-the-Directive-to-the-new-Regulation-2023-07-V1.pdf, Last visited on 15th July 2024.
9. EU Cloud Services Scheme (EUCS), ENISA, available at https://ec.europa.eu/newsroom/cipr/items/713799/en.
10. Regulation (EU) 2019/2144 of the European Parliament and of the Council of 27 November 2019 on type-approval requirements for motor vehicles and their trailers, and systems, components and separate technical units intended for such vehicles, as regards their general safety and the protection of vehicle occupants and vulnerable road users, amending Regulation (EU) 2018/858 of the European Parliament and of the Council and repealing Regulations (EC) No 78/2009, (EC) No 79/2009 and (EC) No 661/2009 of the European Parliament and of the Council and Commission Regulations (EC) No 631/2009, (EU) No 406/2010, (EU) No 672/2010, (EU) No 1003/2010, (EU) No 1005/2010, (EU) No 1008/2010, (EU) No 1009/2010, (EU) No 19/2011, (EU) No 109/2011, (EU) No 458/2011, (EU) No 65/2012, (EU) No 130/2012, (EU) No 347/2012, (EU) No 351/2012, (EU) No 1230/2012 and (EU) 2015/166 (Text with EEA relevance), available at https://eur-lex.europa.eu/eli/reg/2019/2144/oj.
11. Regulation (EU) 2017/745 of the European Parliament and of the Council of 5 April 2017 on medical devices, amending Directive 2001/83/EC, Regulation (EC) No 178/2002 and Regulation (EC) No 1223/2009 and repealing Council Directives 90/385/EEC and 93/42/EEC (Text with EEA relevance), available at https://eur-lex.europa.eu/legal-content/EN/TXT/?uri=celex%3A32017R0745.
12. Ceramic World Web. "New Machinery Regulation: regulatory and safety obligations," January 26, 2024. https://www.ceramicworldweb.com/en/technology/new-machinery-regulation-regulatory-and-safety-obligations Last visited on 16th July 2024.
13. Shen, Yvette. "Machinery Regulation: An Essential Guide." Compliance Gate, February 19, 2024. https://www.compliancegate.com/machinery-regulation/#Conformity_Assessment_Procedures, last visited on 16th July 2024.
14. Machinery Directive, European Commission, available at https://single-market-economy.ec.europa.eu/single-market/european-standards/harmonised-standards/machinery-md_en.

15. Kielesińska, Agata, and Miroslav Pristavka. "The Machinery Safety Management - Selected Issues" System Safety: Human - Technical Facility - Environment, vol. 1, no. 1, Sciendo, 2019, pp. 45–52. https://doi.org/10.2478/czoto-2019-0006.
16. Regulation (EU) 2019/881 of the European Parliament and of the Council of 17 April 2019 on ENISA (the European Union Agency for Cybersecurity) and on information and communications technology cybersecurity certification and repealing Regulation (EU) No 526/2013 (Cybersecurity Act) (Text with EEA relevance).

Chapter 10
Introduction to the European Cybersecurity Act

Markus Mueck and Christophe Gaie

Abstract The 2019 European Cybersecurity Act renews and redefines the mandate of the European Union Agency for Cybersecurity (ENISA) which was originally created in 2004 as the European Network and Information Security Agency. The objective is to ensure a high level of cybersecurity, cyber resilience and trust across the EU and for this purpose builds on voluntary European cybersecurity certification schemes for ICT (Information and Communications Technology) products, services and processes. In 2024, the European Commission had issued a Delegated Act related to the adoption of the voluntary European common criteria-based cybersecurity certification scheme (EUCC)—the first scheme adopted at EU level. It introduces certificates for ICT products (i.e., hardware and software including components such as chips and smart cards) at the 'substantial' or 'high' assurance levels. Further cybersecurity certification schemes in the pipeline are the EU Cloud Services Scheme (EUCS) and 5G Cybersecurity Certification Scheme. In the present chapter, we will give a detailed summary of the Cybersecurity Act and ENISA's role as established therein. Also, a summary of the above-mentioned Cybersecurity Certification Schemes will be given which constitute an additional pillar in addition to other Cybersecurity initiatives including the Radio Equipment Directive (RED) Articles 3(3)(d/e/f) on Cybersecurity and Privacy and the novel Cyber Resilience Act (CRA). Finally, interdependencies with other regulations are being discussed as it is for example the case for the renewed European Machinery Directive.

Keywords Cybersecurity · Regulation · European Union single market · Risk assessment · Harmonized standards

M. Mueck (✉)
Intel Deutschland GmbH, Munich, Germany
e-mail: Markus.Dominik.Mueck@intel.com

C. Gaie
Head of the Engineering and Digital Innovation Division, Prime Minister Services, Paris, France
e-mail: christophe.gaie@gmail.com

M. Mueck and C. Gaie (eds.), *European Digital Regulations*, Intelligent Systems Reference Library 265, https://doi.org/10.1007/978-3-031-80809-8_10

10.1 Introduction

In 2004, EU Regulation 460/2004 [1] originally created ENISA as the European Network and Information Security Agency. In 2008, the European Parliament and the Council adopted Regulation (EC) No 1007/2008 [2] extending the mandate of the Agency until March 2012. Then, Regulation (EC) No 580/2011 [3] further extended the mandate of the Agency until 13 September 2013. Subsequently, Regulation (EC) No 526/2013 [4] transformed ENISA into the European Union Agency for Network and Information Security. This Regulation was succeeded by the European Cybersecurity Act [5] which renews and redefines ENISA role, now entitled European Union Agency for Cybersecurity. ENISA is established for an indefinite period as from 27 June 2019.

As stated in the European Cybersecurity Act (Article 17 [5]), ENISA is mandated to perform various cybersecurity tasks. These include providing advice and expertise, serving as a Union information hub, fostering best practice exchange, offering policy suggestions to the Commission and Member States, acting as a **cybersecurity reference point** for Union sectoral policy initiatives, and promoting operational cooperation among Member States and Union institutions.

As a basic principle, the European Cybersecurity Act introduces three Cybersecurity Assurance levels as follows (Fig. 10.1).

As stated in the European Cybersecurity Act (Article 52 [5]), the assurance level should match the risk of the ICT (Information and Communications Technology) product, service, or process, considering the likelihood and impact of a security incident.

Furthermore, Article 52 of the European Cybersecurity Act [5] introduces specific objectives for those three Assurance Levels, which noting that it is up to the European Cybersecurity Certification Schemes to specify one or more Assurance Levels for ICT products, ICT services and ICT processes (Fig. 10.2).

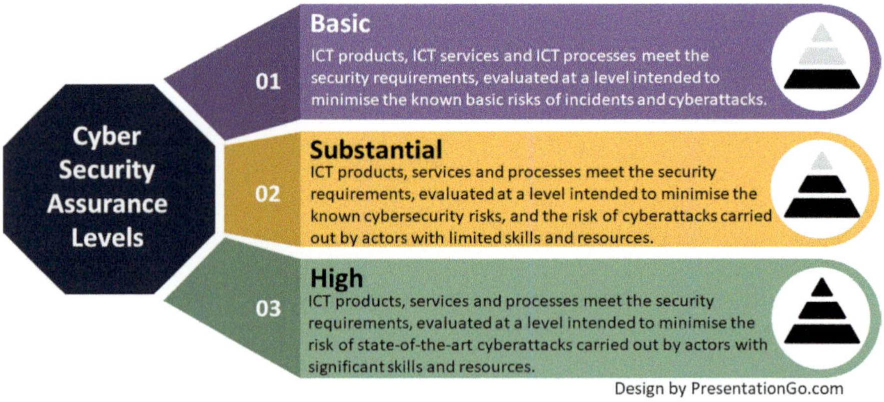

Fig. 10.1 Three cybersecurity assurance levels as defined by the European Cybersecurity Act [5]

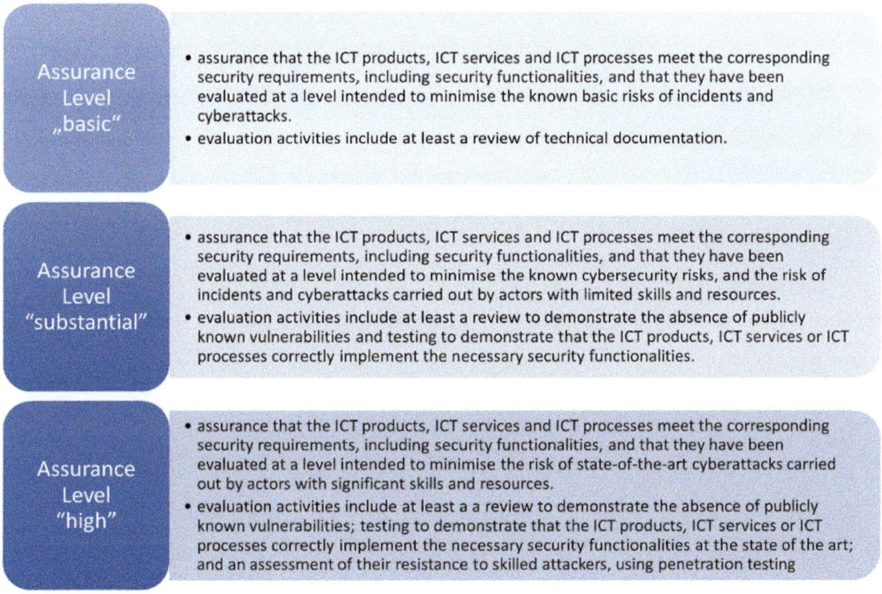

Fig. 10.2 Implementation of the three cybersecurity assurance levels as defined by the European Cybersecurity Act [5]

The European Cybersecurity Act [5] furthermore introduces a European Cybersecurity Certification Framework, laying down the requirements for European Cybersecurity Certification Schemes ([5], point 69). This framework will enable European Cybersecurity Certificates and EU Statements of Conformity for ICT products, services, or processes to be recognized and utilized across all Member States.

In 2024, the European Commission has adopted the first-ever European cybersecurity certification scheme through Commission Implementing Regulation (EU) 2024/482 [6], in line with the EU Cybersecurity Act: European Common Criteria-based cybersecurity certification scheme (EUCC). As indicated by the European Commission [7], the voluntary scheme will supplement the Cyber Resilience Act, which imposes mandatory cybersecurity standards on all hardware and software products within the EU.

The EUCC Implementing Regulation will apply as of 27 February 2025. We note that the application of the schemes remains voluntary—in practice, however, they may become quasi-mandatory for manufacturers in case that customers require compliance to those schemes. Also, it should be monitored whether the application may become mandatory in the future on the level of Member States or the European Union as a whole.

Further developing Cybersecurity Certifications schemes are the European Certification Scheme for Cloud Services (EUCS) and the European Cybersecurity Certification Scheme for 5G as detailed by ENISA [8]. Those will further be detailed within the present chapter and are illustrated in Fig. 10.3.

Fig. 10.3 Cybersecurity certification schemes (available or developing)

ENISA is currently evaluating the need for a Cybersecurity Certification scheme in light of the European AI Act [9]. Additionally, as noted in [8], Managed Security Services are pivotal in preventing and responding to Cybersecurity threats and incidents, as emphasized in the NIS2 Directive and hinted at in the upcoming Cyber Solidarity Act. The Union Rolling Work Programme for Certification already includes these services, and the future amendment of the Cybersecurity Act is expected to authorize ENISA to certify them. The European Agency for Cybersecurity has initiated preparatory work on this matter.

The 2024 Union Rolling Work Programme for European cybersecurity certification [10] further outlines areas for possible future European cybersecurity certification schemes as summarized in Fig. 10.4.

Note that the summaries given above are based on the information available at the time when the present chapter was written. Obviously, the plans of ENISA are

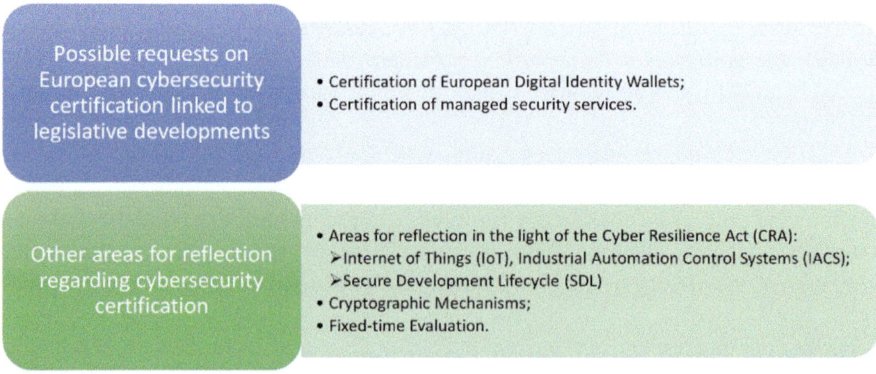

Fig. 10.4 Further areas for possible future European cybersecurity certification schemes [10]

likely to evolve over time and thus the areas for possible future European cybersecurity certification schemes may follow. Also, in addition to the upper information, Legislations [11–13] should be considered.

10.2 European Union Agency for Cybersecurity (ENISA)

The sequence of Regulations which lead to the creation of the European Union Agency for Cybersecurity (ENISA) as explained above. ENISA's mandate, tasks and administrative and management structure are summarized below as outlined in [14] (Fig. 10.5).

In the upper structure, ENISA's Management Board is composed of one representative from each EU Member State and two members appointed by the European Commission. The Executive Board is composed of five members.

The ENISA Advisory Group is composed of recognised experts from relevant stakeholders, such as the ICT (Information and Communications Technology) industry, organizations which provide electronic communications networks or services, SMEs (small and medium-sized enterprises), consumers, representatives of academia and operators of essential services, representatives of competent authorities

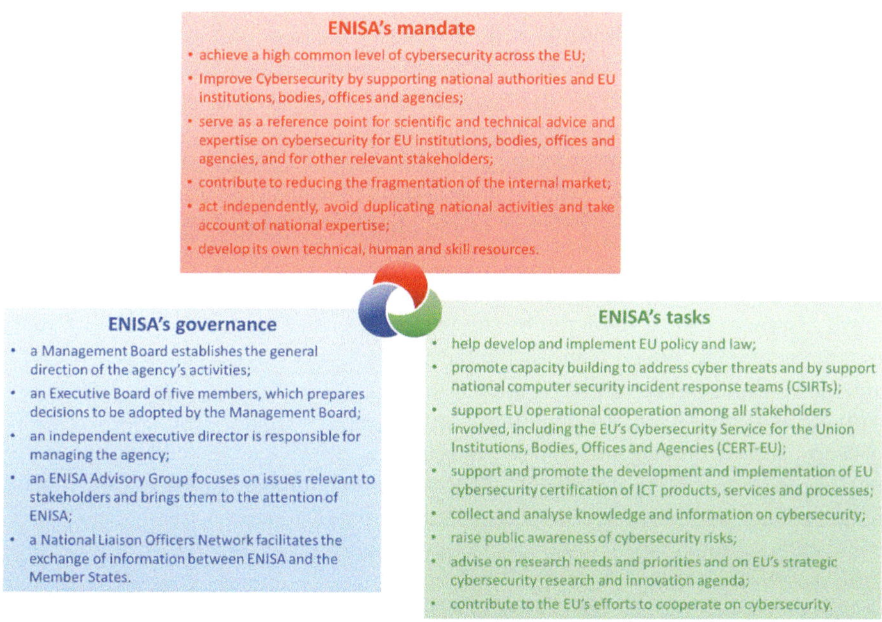

Fig. 10.5 ENISA's mandate, tasks and governance (shortened) [14]

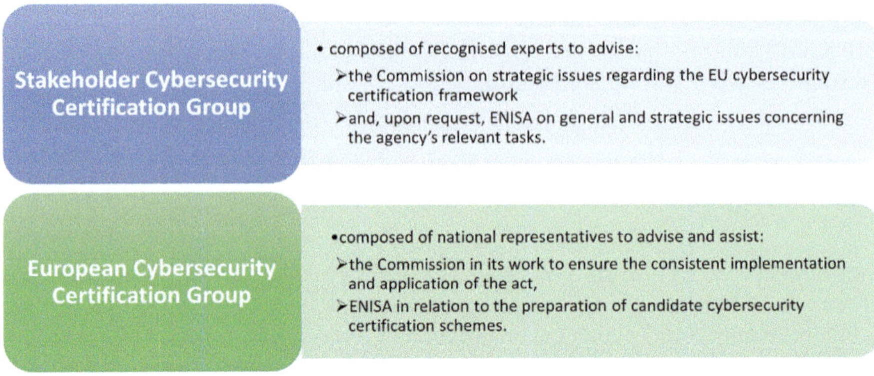

Fig. 10.6 Groups established under the Cybersecurity Act

(as notified under the EECC (European Electronic Communications Code)), standardisation organisations, law enforcement and data protection supervisory authorities. To conclude, representatives of all Member States constitute the National Liaison Officers Network.

The Cybersecurity Act furthermore established the following groups (Fig. 10.6).

In the authors' opinion, ENISA will provide many benefits. First, it serves as a central hub for cybersecurity expertise within the EU, offering guidance and support to member states and organizations. By promoting standardized cybersecurity practices and reducing fragmentation, ENISA enhances interoperability across Europe. Moreover, the agency collects and analyzes threat intelligence, providing valuable insights to stakeholders for risk mitigation. Additionally, ENISA assists in developing cybersecurity capabilities within member states, particularly those with limited resources. The agency also fosters collaboration between public and private sector entities, facilitating information sharing and joint initiatives to address cybersecurity challenges effectively.

However, the authors also identify some limitations concerning ENISA. Indeed, its effectiveness can be hindered by the willingness and capacity of member states to implement its recommendations. Moreover, the ENISA agency may face resource constraints that limit its ability to address all cybersecurity challenges. The rapidly evolving and complex nature of cybersecurity threats can make it difficult for ENISA to keep pace with emerging issues. Additionally, ENISA's mandate and authority may be limited, making it difficult to enforce its recommendations or directives in certain situations. Finally, the varying cybersecurity challenges across member states due to cultural and political differences can pose difficulties for ENISA in addressing these issues effectively.

10.3 The European Common Criteria-Based Cybersecurity Certification Scheme (EUCC)

The European Common Criteria-based Cybersecurity Certification Scheme (EUCC) is the first-ever European cybersecurity certification scheme. A corresponding Commission Implementing Regulation (EU) 2024/482 [6] was issued in 2024, in line with the EU Cybersecurity Act. As indicated by the European Commission [7], the voluntary scheme will supplement the Cyber Resilience Act, which imposes mandatory cybersecurity standards on all hardware and software products within the EU. The EUCC Implementing Regulation will take effect on February 27, 2025.

The European Common Criteria-based Cybersecurity Certification Scheme (EUCC) is concerned with the certification of ICT (Information and Communications Technology) products cybersecurity; it is based on the Common Criteria, the Common Methodology for Information Technology Security Evaluation, and corresponding standards, respectively, as defined by ISO/IEC 15408 [15] and ISO/IEC 18045 [16]. Based on the CSA Article 49.1, the European Common Criteria-based Cybersecurity Certification Scheme (EUCC) addresses the necessary requirements associated with the definition of a scheme, prescribing that the requirements of Articles 51, 52 and 54 of the CSA shall be met.

10.3.1 Implementing Regulation for EUCC

The Commission Implementing Regulation (EU) 2024/482 [6] was published in 2024 to establish guidelines for implementing Regulation (EU) 2019/881 of the European Parliament and of the Council regarding the adoption of the European Common Criteria-based cybersecurity certification scheme (EUCC).

Article 1 of the Commission Implementing Regulation (EU) 2024/482 [6] states that the regulation applies to all ICT (Information and Communications Technology) products, including their documentation, submitted for certification under the European Common Criteria-based Cybersecurity Certification Scheme (EUCC), and to all protection profiles submitted for certification as part of the ICT process leading to ICT product certification. Certification bodies are authorized to issue EUCC certificates at 'substantial' or 'high' assurance levels. Self-assessment of conformity is not allowed. The validity period of EUCC certificates cannot exceed 5 years.

10.3.2 Summary of the European Common Criteria-Based Cybersecurity Certification Scheme (EUCC)

The European Commission outlines the European Common Criteria-based Cybersecurity Certification Scheme (EUCC) in [17]. Notably, as a voluntary initiative, the EUCC scheme empowers ICT suppliers seeking to demonstrate assurance to undergo a standardized EU assessment process for certifying ICT (Information and Communications Technology) products, including technological components (chips, smartcards), hardware, and software.

Furthermore, the European Common Criteria-based cybersecurity certification scheme (EUCC) builds on the MRA (Mutual Recognition Agreement) of Information Technology Security Certificates [18] as developed by the Senior Officials Group Information Systems Security ('SOG-IS') [19]. This approach is relying on the Common Criteria, including the group's procedures and documents [20].

This approach aims to facilitate a seamless transition from national SOG-IS schemes to the unified European Common Criteria-based Cybersecurity Certification Scheme (EUCC) scheme. The European Common Criteria-based Cybersecurity Certification Scheme (EUCC) employs components 1 to 5 of the Common Criteria's vulnerability assessment family (AVA_VAN). To ensure a high level of trust and confidence in certified ICT products, self-assessment will not be permitted under this Regulation. Only third-party conformity assessments conducted by ITSEF (IT Security Evaluation Facilities) and certification bodies will be allowed [6].

To support its implementation, ENISA additionally publishes so called 'state-of-the-art documents' which are documents which specify evaluation methods, techniques and tools that apply to the certification of ICT products or security requirements of a generic ICT product category in order to harmonize evaluation in technical domains or of protection profiles.

Links to the following state-of-the-art documents are available in Annex I of [6] and summarized in [17] (Fig. 10.7).

To obtain the final certification, the concerned IT product is evaluated, at a minimum, with respect to the following criteria (a) to (e) as defined in Article 7 of the Commission Implementing Regulation (EU) 2024/482 [6] (Figs. 10.8 and 10.9).

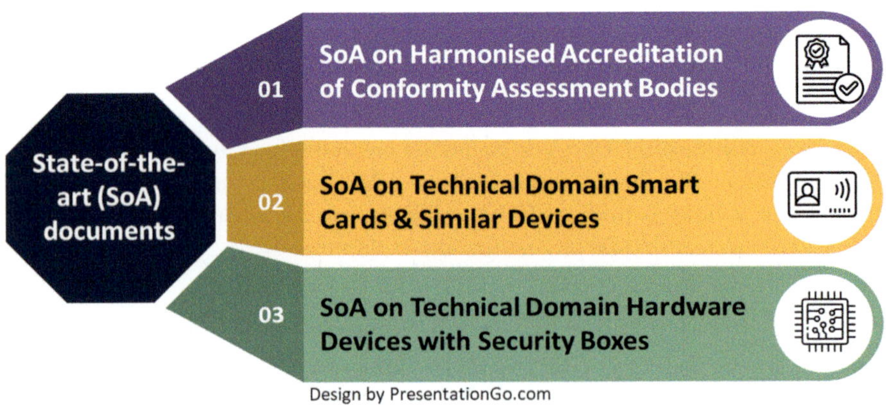

Fig. 10.7 State-of-the-art (SoA) documents in support of the EUCC scheme

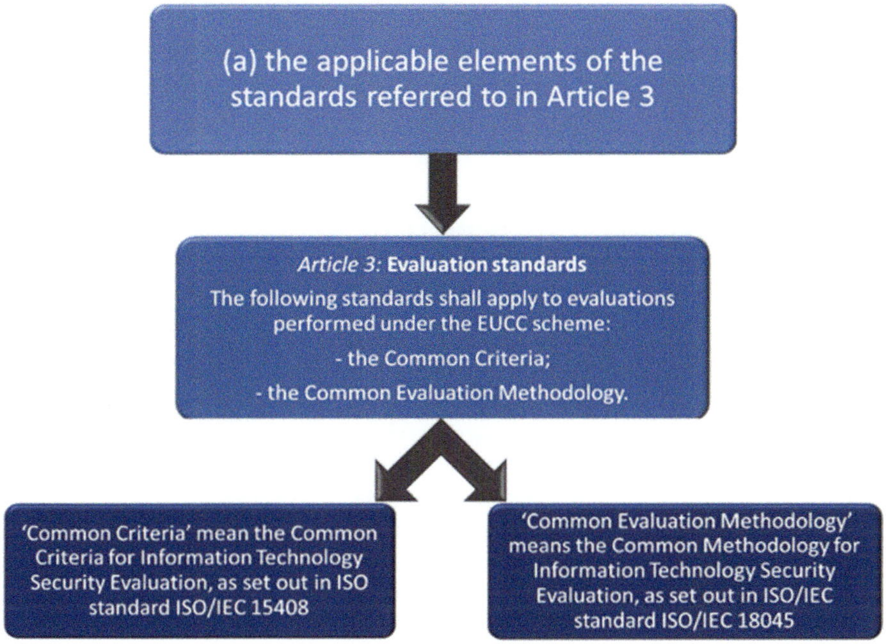

Fig. 10.8 Criterion (**a**) of Article 7 of the Commission Implementing Regulation (EU) 2024/482 [6]

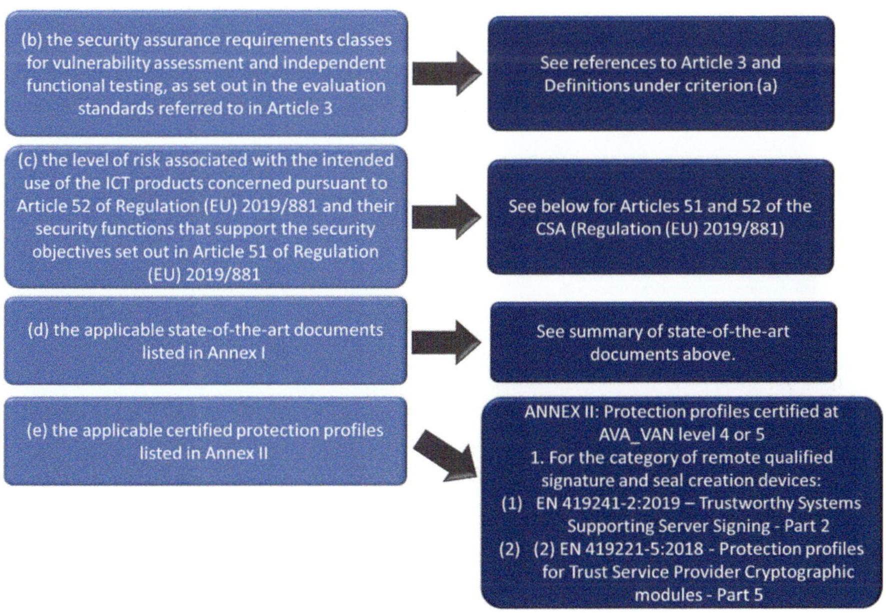

Fig. 10.9 Criteria (**b**, **c**, **d**, **e**) of Article 7 of the Commission Implementing Regulation (EU) 2024/482 [6]

Article 51 the Cybersecurity Act, i.e. Regulation (EU) 2019/881, provides a list of Security objectives of European cybersecurity certification schemes. Article 52 introduces Assurance levels of European cybersecurity certification schemes, notably Assurance levels 'basic', 'substantial' or 'high'.

10.4 EU Cloud Services Scheme (EUCS)

At the time when the present chapter was written, the European Cybersecurity Certification Scheme for Cloud Services (EUCS) was available as a Draft [21] thus considered to be a *Candidate Scheme*. EUCS relates to the certification of the cybersecurity of cloud services. The candidate scheme refers to a specific category of ICT (Information and Communications Technology) services, this is the reason why it relies on ISO/IEC 17065 [22] to determine applicable requirements to CABs (Conformance Assessment Bodies). Furthermore, the ISO 27000 [23] series of standards and on the International Auditing Standards are commonly used for the assessment of the cybersecurity of cloud services.

As outlined in [21], the EUCS scheme seeks to enhance the Internal Market and bolster the security of diverse cloud services and their underlying application, infrastructure, and platform capabilities. The EUCS scheme comprehensively addresses security requirements by offering three levels of assurance defined in the EUCSA: basic, substantial, and high. Potential users of the scheme include (Fig. 10.10).

The EUCS indeed enables CSPs to assess how a cloud service meets the requirements of a predefined set of security control objectives and a related set of measures, to enable CSCs to assess cloud offerings and meet own security compliance requirements and to allow regulatory authorities to engage in compliance verification.

Fig. 10.10 Users of the EUCS scheme

Chapter 8 of the Draft EUCS Scheme [21] introduces Evaluation Methods and Criteria, while Annex C details how to conduct the assessment for levels Substantial and High; Annex D provides details for the assessment for level Basic.

The offerings of the Draft EUCS Scheme [21] is summarized below for CSPs and CSCs (Fig. 10.11).

The scheme relies on a number of standards and technical specifications (Fig. 10.12).

As stated above, the detailed assessment methodology is detailed in Annex C for Assurance Levels Substantial and High and in Annex D for the Assurance Level Basic.

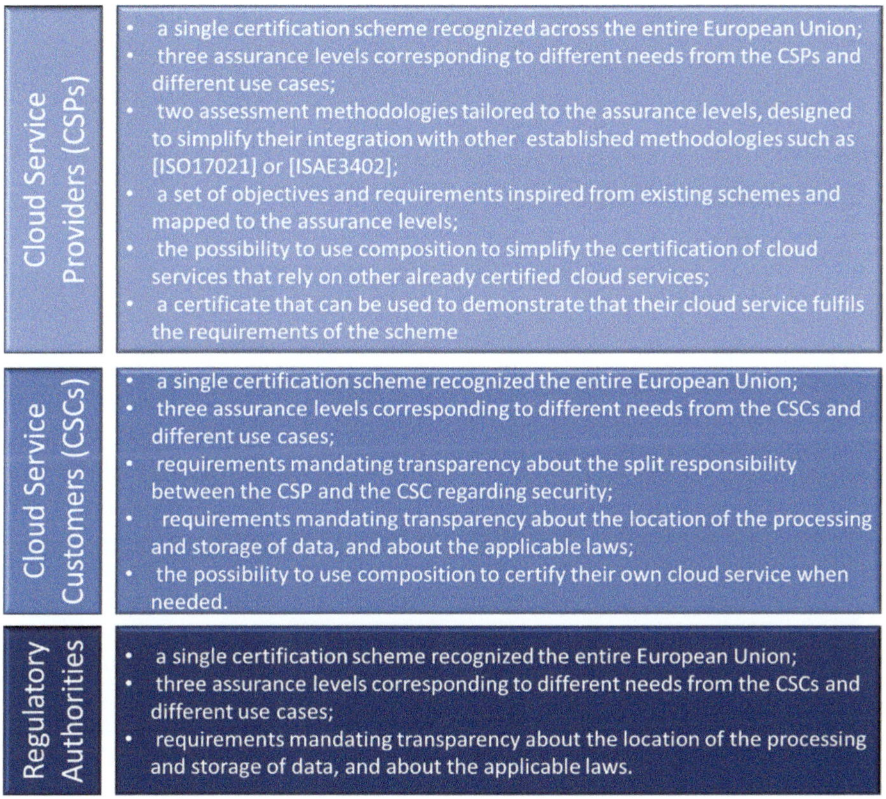

Fig. 10.11 Offerings by the Draft EUCS scheme [21]

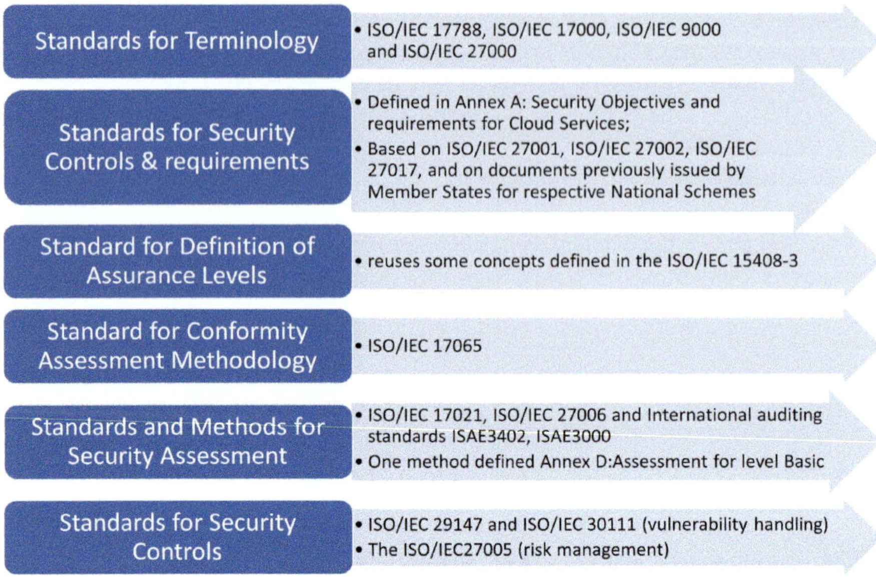

Fig. 10.12 Standards upon which the EUCC scheme is relying

10.5 5G Cybersecurity Certification (EU5G Scheme)

ENISA has published a threat landscape mapping complementing the EU coordinated risk assessment [24]: *ENISA threat landscape for 5G Networks* [25]. Following the EU coordinated risk assessment, ENISA published an EU Toolbox of risk mitigating measures (EU Toolbox for 5G Security [26]). Then, the European Commission requested ENISA to develop a candidate European cybersecurity certification scheme for 5G networks (EU 5G scheme) with the objective to enhance the framework on 5G cybersecurity and secure 5G deployment in the European Union. This is addressed under the EU5G Ad-Hoc Working Group (AHWG) [27].

At the time when the present chapter was written, there was limited public information available on the Draft 5G Cybersecurity Certification Scheme. One source is a presentation by ENISA to the ETSI Cybersecurity Conference in October 2023 [28]. Another source is a November 2022 presentation by ENISA to 3GPP (3rd Generation Partnership Project) [29], followed by an ENISA presentation to the plenary meeting of the 3GPP Technical Specification Group Service and System Aspects (TSG SA) in March 2023 [30]. ENISA presentation [29] includes information on a time line as follows (Fig. 10.13).

ENISA Presentation [30] furthermore discusses the structure of EU5G in terms of Thematic Groups (TGs) as follows (Fig. 10.14).

In the upper illustration, the acronym NESAS refers to the Global System for Mobile Communications Association (GSMA) Network Equipment Security Assurance Scheme (NESAS) [31]. It is the typical Security Assurance Scheme applied

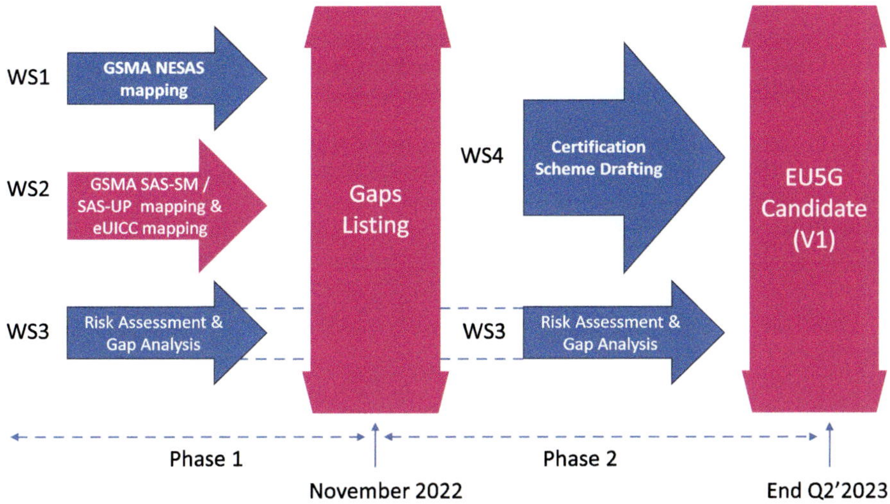

Fig. 10.13 EU5G AHWG milestones [29]

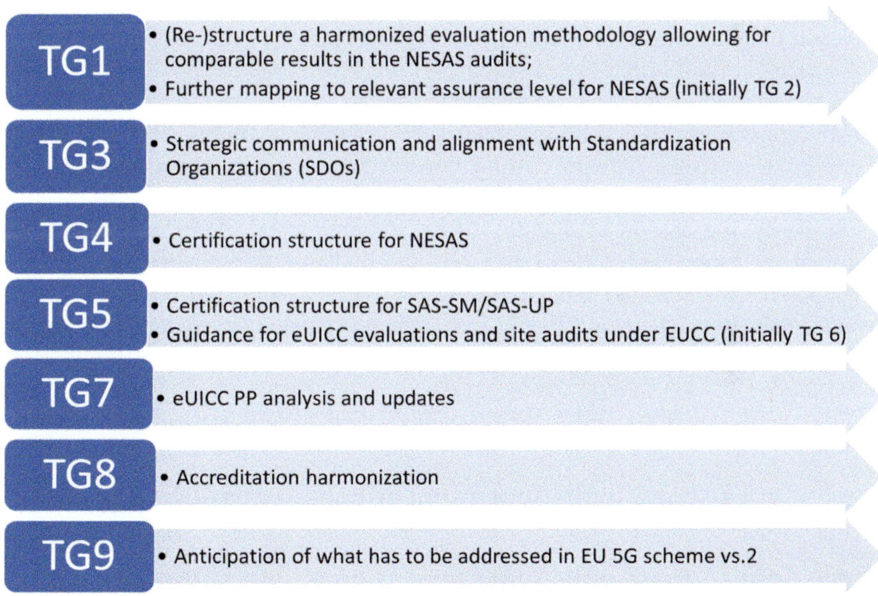

Fig. 10.14 Structure of EU5G, thematic groups (TGs) [30]

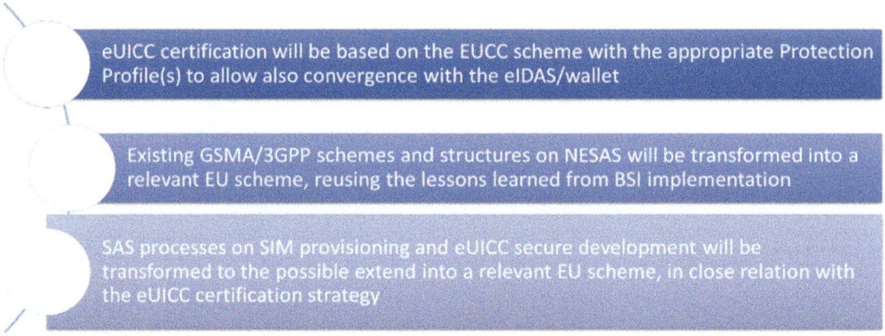

Fig. 10.15 Three different components of the EU5G scheme [33]

by manufacturers of cellular network equipment. The acronym SAS relates to the GSMA Security Accreditation Scheme [32]. SAS-UP is the SAS for UICC Production. SAS-SM is the SAS for Subscription Management. UICC is the Universal Integrated Circuit Card, i.e. the smart card used in mobile terminals; the UICC contains a Subscriber Identification Module (SIM) application. eUICC is the Embedded Universal Integrated Circuit Card. eUICC PP is the eUICC Protection Profile.

As stated in the ENISA presentation [33], the EU5G Scheme will be organized into 3 different components as illustrated in Fig. 10.15.

In the upper illustration, BSI is the German Federal Office for Information Security which has developed a national certification scheme for product certification based on the GSMA-NESAS assessment scheme [34].

Obviously, the date of availability of a Candidate EU5G Certification Scheme is shifted. In June 2024, ENISA had released and was seeking feedback on the embedded Universal Integrated Circuit Card (eUICC) specifications of the cybersecurity certification scheme on EU5G [35], which is carried out under the Common Criteria scheme [36]. These specifications are part of the ENISA work under the EU5G AHWG.

10.6 Interdependencies with Other European Regulations

10.6.1 Cyber Resilience Act (CRA)

The Cyber Resilience Act (CRA), which was available as a Draft [37] (see [38, 39] for earlier draft versions) at the time when the present chapter was written, has some complex interdependencies with the Cybersecurity Act [5].

The draft Cyber Resilience Act clearly aims to align with the Cybersecurity Act, Regulation (EU) 2019/881. This regulation establishes a voluntary European cybersecurity certification framework for ICT products, processes, and services. European cybersecurity certification schemes offer a standardized trust framework for users to utilize products with digital components covered by this regulation. Therefore, the Cyber Resilience Act should effectively integrate with Regulation (EU) 2019/881.

Specifically, for critical products as defined by the CRA [37], the EU Cybersecurity Certification Schemes are referenced: … *potential mandating of certification under a European cybersecurity certification scheme of the critical products with digital elements set out in an annex to this Regulation, as well as for updating the list of critical products with digital elements based on criticality criteria set out in this Regulation, and for specifying the European cybersecurity certification schemes adopted pursuant to Regulation (EU) 2019/881 that can be used to demonstrate conformity with the essential cybersecurity requirements or parts thereof as set out in an annex to this Regulation.*"

10.6.2 Artificial Intelligence Act (AI Act)

Regarding Cybersecurity, the European Artificial Intelligence Act (AI Act) [9] explicitly mentions the European Cybersecurity Act, Regulation (EU) 2019/881 [5]. The AI Act recognizes interdependencies and stipulates that the European Parliament and Council (37), along with the Commission, should collaborate with ENISA on matters concerning the cybersecurity of AI systems. Moreover, Article 42 of the AI Act addresses the presumption of conformity with certain requirements.

High-risk AI systems that have been certified or for which a statement of conformity has been issued under a cybersecurity scheme pursuant to Regulation (EU) 2019/881 (…) shall be presumed to comply with the cybersecurity requirements set out in Article 15 (…).

Article 15 of the AI Act specifically relates to ***accuracy, robustness and cybersecurity***.

10.6.3 Machinery Regulation

In terms of Cybersecurity, the Machinery Regulation [40] specifically references the European Cybersecurity Act (Regulation (EU) 2019/881) [5]. The Machinery Regulation stipulates that, to address risks arising from malicious third-party actions that compromise product safety within its scope, it should incorporate essential health and safety requirements. Compliance with these requirements can be assumed based on a certificate or statement of conformity issued under a relevant cybersecurity certification scheme adopted in accordance with Regulation (EU) 2019/881 of the European Parliament and of the Council.

Cybersecurity Certification Schemes are being developed under the Cybersecurity Act [5] under control of the European Union Agency for Cybersecurity (ENISA) as explained in the previous sections of the present chapter.

The Machinery Regulation [40] elaborates on how presumption of conformity can be established through a valid Cybersecurity Certification Scheme under the European Cybersecurity Act (Regulation (EU) 2019/881) [5]: Machinery and related products that have undergone certification or received a statement of conformity under a cybersecurity certification scheme adopted in compliance with Regulation (EU) 2019/881, whose references have been published in the Official Journal of the European Union, shall be deemed to comply with the essential health and safety requirements outlined in Annex III, sections 1.1.9 and 1.2.1, concerning protection against corruption and the safety and reliability of control systems, to the extent that these requirements are covered by the cybersecurity certificate or statement of conformity or relevant portions thereof.

10.6.4 The Networks and Information Systems 2 (NIS2) Directive

The NIS2 Directive [41], in its Article 24, references the European Cybersecurity Act [5] regarding the use of European cybersecurity certification schemes. Under NIS2 Article 21, Member States may require essential and important entities to utilize specific ICT products, services, or processes, either developed internally or procured from third parties, that are certified under European cybersecurity certification schemes established by the European Cybersecurity Act [5]. Moreover, if inadequate cybersecurity levels are detected, the European Commission has the authority to mandate the use of certain certified ICT products, services, or processes or require entities to obtain a certificate under a European cybersecurity certification scheme through Delegated Acts.

10.7 Conclusion

The Cybersecurity Act (CSA) is a major European Regulation. Among other items, it establishes the European Union Agency for Cybersecurity (ENISA) and introduces the concept of **EU Cybersecurity Certification Schemes**. When the present chapter was written, three of those EU Cybersecurity Certification Schemes were either available or under development. Further Schemes are expected to follow which will constitute a major pillar in the European Cybersecurity strategy. Also, a number of other Regulations rely on the EU Cybersecurity Certification Schemes for conformity assessment.

Further Cybersecurity related Legislations, including the Cyber Resilience Act (CRA) [38, 39] and the Network Information Security Directive (NIS2) [41]. It is thus important for stakeholders to carefully monitor the further evolution in this space. Moreover, further research should describe the concrete implementations of EU Cybersecurity Certification Schemes in various case studies such as the energy sector [42], or other digitalized domains.

References

1. Regulation (EC) No 460/2004 of the European Parliament and of the Council of 10 March 2004 establishing the European Network and Information Security Agency (Text with EEA relevance), available at https://eur-lex.europa.eu/legal-content/EN/TXT/?uri=celex%3A32004R0460.
2. Regulation (EC) No 1007/2008 of the European Parliament and of the Council of 24 September 2008 amending Regulation (EC) No 460/2004 establishing the European Network and Information Security Agency as regards its duration (Text with EEA relevance), available at https://eur-lex.europa.eu/legal-content/EN/TXT/?uri=CELEX%3A32008R1007.
3. Regulation (EU) No 580/2011 of the European Parliament and of the Council of 8 June 2011 amending Regulation (EC) No 460/2004 establishing the European Network and Information Security Agency as regards its duration Text with EEA relevance, available at https://eur-lex.europa.eu/legal-content/EN/ALL/?uri=CELEX%3A32011R0580.
4. Regulation (EU) No 526/2013 of the European Parliament and of the Council of 21 May 2013 concerning the European Union Agency for Network and Information Security (ENISA) and repealing Regulation (EC) No 460/2004 Text with EEA relevance, available at https://eur-lex.europa.eu/eli/reg/2013/526/oj.
5. Regulation (EU) 2019/881 of the European Parliament and of the Council of 17 April 2019 on ENISA (the European Union Agency for Cybersecurity) and on information and communications technology cybersecurity certification and repealing Regulation (EU) No 526/2013 (Cybersecurity Act) (Text with EEA relevance), available at https://eur-lex.europa.eu/legal-content/EN/TXT/?uri=celex%3A32019R0881.
6. Commission Implementing Regulation (EU) 2024/482 of 31 January 2024 laying down rules for the application of Regulation (EU) 2019/881 of the European Parliament and of the Council as regards the adoption of the European Common Criteria-based cybersecurity certification scheme (EUCC) (OJ L, 2024/482, 7.2.2024), available at https://eur-lex.europa.eu/eli/reg_impl/2024/482/oj.
7. First EU-wide cybersecurity certification scheme to make European digital space safer, Press release, 31 January 2024, available at https://digital-strategy.ec.europa.eu/en/news/first-eu-wide-cybersecurity-certification-scheme-make-european-digital-space-safer.
8. Developing Certification Schemes, ENISA, available at https://certification.enisa.europa.eu/about-eu-certification/developing-certification-schemes_en.
9. REGULATION (EU) 2024/1689 OF THE EUROPEAN PARLIAMENT AND OF THE COUNCIL of 13 June 2024 laying down harmonised rules on artificial intelligence and amending Regulations (EC) No 300/2008, (EU) No 167/2013, (EU) No 168/2013, (EU) 2018/858, (EU) 2018/1139 and (EU) 2019/2144 and Directives 2014/90/EU, (EU) 2016/797 and (EU) 2020/1828 (Artificial Intelligence Act), available at https://eur-lex.europa.eu/eli/reg/2024/1689/oj.
10. COMMISSION STAFF WORKING DOCUMENT: Union Rolling Work Programme for European cybersecurity certification, February 2024, available at https://digital-strategy.ec.europa.eu/en/library/union-rolling-work-programme-european-cybersecurity-certification.

11. Regulation (EU) 2018/1725 of the European Parliament and of the Council of 23 October 2018 on the protection of natural persons with regard to the processing of personal data by the Union institutions, bodies, offices and agencies and on the free movement of such data, and repealing Regulation (EC) No 45/2001 and Decision No 1247/2002/EC (OJ L 295, 21.11.2018, pp. 39–98).
12. Directive (EU) 2016/1148 of the European Parliament and of the Council of 6 July 2016 concerning measures for a high common level of security of network and information systems across the Union (OJ L 194, 19.7.2016, pp. 1–30).
13. Regulation (EU) 2016/679 of the European Parliament and of the Council of 27 April 2016 on the protection of natural persons with regard to the processing of personal data and on the free movement of such data, and repealing Directive 95/46/EC (General Data Protection Regulation) (OJ L 119, 4.5.2016, pp. 1–88).
14. Summaries of EU Legislation: The Cybersecurity Act, European Commission, available at https://eur-lex.europa.eu/EN/legal-content/summary/the-eu-cybersecurity-act.html.
15. ISO/IEC 15408–1:2022, Information security, cybersecurity and privacy protection — Evaluation criteria for IT security, Part 1: Introduction and general model, available at https://www.iso.org/standard/72891.html.
16. ISO/IEC 18045:2022, Information security, cybersecurity and privacy protection — Evaluation criteria for IT security — Methodology for IT security evaluation, available at https://www.iso.org/standard/72889.html.
17. EUCC Certification Scheme, European Commission, available at https://certification.enisa.europa.eu/certification-library/eucc-certification-scheme_en.
18. Mutual Recognition Agreement of Information Technology Security Evaluation Certificates, VERSION 3.0, available at https://www.sogis.eu/uk/mra_en.html.
19. Senior Officials Group Information Systems Security ('SOG-IS'), available at https://www.sogis.eu/.
20. Implementing Regulation on the adoption of a European Common Criteria-based cybersecurity certification scheme, European Commission, January 2024, available at https://digital-strategy.ec.europa.eu/en/library/implementing-regulation-adoption-european-common-criteria-based-cybersecurity-certification-scheme.
21. Draft EUCS – Cloud Services Scheme, ENISA, available at https://www.enisa.europa.eu/publications/eucs-cloud-service-scheme.
22. ISO/IEC 17065:2012, Conformity assessment — Requirements for bodies certifying products, processes and services, available at https://www.iso.org/standard/46568.html.
23. ISO/IEC 27000:2018, Information technology — Security techniques — Information security management systems — Overview and vocabulary, available at https://www.iso.org/standard/73906.html.
24. EU coordinated risk assessment of the cybersecurity of 5G networks, European Commission and the European Agency for Cybersecurity, 2019, available at https://ec.europa.eu/newsroom/dae/document.cfm?doc_id=62132.
25. ENISA threat landscape for 5G Networks, ENISA, available at https://www.enisa.europa.eu/publications/enisa-threat-landscape-for-5g-networks.
26. The EU toolbox for 5G security, ENISA; available at https://digital-strategy.ec.europa.eu/en/library/eu-toolbox-5g-security.
27. Ad-Hoc Working Group 03 - on 5G Cybersecurity Certification, ENISA, available at https://www.enisa.europa.eu/topics/certification/copy_of_adhoc_wg_calls/ad-hoc-working-group-on-5g-cybersecurity-certification.
28. DEVELOPING AN EU 5G CERTIFICATION SCHEME presented at ETSI Security Conference, ENISA, October 2023, available at https://docbox.etsi.org/Workshop/2023/10_ETSISECURITYCONFERENCE/D13_REGULATION/ENISA_MAGNABOSCO.pdf.
29. Cooperation on Cybersecurity, 3GPP, available at https://www.3gpp.org/news-events/partner-news/sa3-enisa.
30. EU 5G Scheme Development, ENISA, 3GPP SA Plenary, Rotterdam, March 2023, available at https://www.3gpp.org/ftp/TSG_SA/TSG_SA/TSGS_99_Rotterdam_2023-03/INBOX/DRAFTS/20230322-EU5G-3GPPSA-final.pptx.

31. Network Equipment Security Assurance Scheme (NESAS), GSMA, available at https://www.gsma.com/solutions-and-impact/industry-services/assurance-services/network-equipment-security-assurance-scheme-nesas/.
32. SAS Accredited Sites, GSMA, available at https://www.gsma.com/solutions-and-impact/industry-services/assurance-services/security-accreditation-scheme-sas/sas-accredited-sites/.
33. AN UPDATE ON SCHEMES DEVELOPMENTS, ENISA, March 2023, available at https://eucyberact.org/wp-content/uploads/2023/03/B12a-BlotP.pdf.
34. Certification according to NESAS CCS-GI, German Federal Office for Information Security, available at https://www.bsi.bund.de/EN/Themen/Unternehmen-und-Organisationen/Standards-und-Zertifizierung/Zertifizierung-und-Anerkennung/Zertifizierung-von-Produkten/Zertifizierung-nach-NESAS/NESAS-CCS-GI_node.html.
35. Specifications related to the certification of the embedded Universal Integrated Circuit Card, ENISA, available at https://certification.enisa.europa.eu/publications/specifications-related-certification-embedded-universal-integrated-circuit-card_en.
36. Provide your Feedback: ENISA public consultation bolsters EU5G Cybersecurity Certification, ENISA, https://certification.enisa.europa.eu/news/provide-your-feedback-enisa-public-consultation-bolsters-eu5g-cybersecurity-certification-2024-06-26_en.
37. REGULATION OF THE EUROPEAN PARLIAMENT AND OF THE COUNCIL on horizontal cybersecurity requirements for products with digital elements and amending Regulations (EU) No 168/2013 and (EU) 2019/1020 and Directive (EU) 2020/1828 (Cyber Resilience Act), 25 September 2024, version adopted by the European Council on 10 October 2024, available at https://data.consilium.europa.eu/doc/document/PE-100-2023-INIT/en/pdf.
38. Proposal for a REGULATION OF THE EUROPEAN PARLIAMENT AND OF THE COUNCIL on horizontal cybersecurity requirements for products with digital elements and amending Regulation (EU) 2019/1020, 15th September 2022, available at https://eur-lex.europa.eu/legal-content/EN/TXT/?uri=celex%3A52022PC0454.
39. final compromise for a REGULATION OF THE EUROPEAN PARLIAMENT AND OF THE COUNCIL on horizontal cybersecurity requirements for products with digital elements and amending Regulation (EU) 2019/1020, 20th December 2023, available at https://eur-lex.europa.eu/legal-content/EN/TXT/PDF/?uri=CONSIL:ST_17000_2023_INIT.
40. "Regulation (EU) 2023/1230 of the European Parliament and of the Council of 14 June 2023 on machinery and repealing Directive 2006/42/EC of the European Parliament and of the Council and Council Directive 73/361/EEC", Last visited on 15th July 2024, available at https://eur-lex.europa.eu/eli/reg/2023/1230/oj..
41. Directive (EU) 2022/2555 of 14 December 2022 on measures for a high common level of cybersecurity across the Union, amending Regulation (EU) No 910/2014 and Directive (EU) 2018/1972, and repealing Directive (EU) 2016/1148 (NIS 2 Directive) [2022] OJ L333/80, available at https://eur-lex.europa.eu/eli/dir/2022/2555/oj.
42. Nwankwo I, Stauch M, Radoglou-Grammatikis P, Sarigiannidis P, Lazaridis G, Drosou A, Tzovaras D. Data Protection and Cybersecurity Certification Activities and Schemes in the Energy Sector. Electronics. 2022; 11(6):965. https://doi.org/10.3390/electronics11060965.